LIFEBOAT STATION DEVELOPMENT PROJECT
ENGINEERING DESIGN REPORT

London South Bank University

Bogdan Ciocoiu and Laurel Milwid (2016)

PURBECK HERITAGE COAST, DORSET

THE DESIGN TEAM

At **Ciocoiu+Milwid**, we are a small team, and therefore each of us has been involved in the range of tasks associated with the design programme.

We operate on a flexible and democratic basis, and we allocate control of tasks to the member whose skill-set allows it to be accomplished most efficiently.

While Bogdan is incredibly efficient at project-planning, and Laurel has strong analytic skills, both have enjoyed the challenge of developing the design for this lifeboat station and applying their creative skills to the design process.

OUR DESIGN ADVISORS

Stephen Vary	Senior Lecturer, School of the Built Environment & Architecture, London South Bank University
David Rose	David Rose Associates
Nick Maclean	Ecos Maclean

We would like to place on record our appreciation to our team of advisors, who guided us so astutely, creatively, and encouragingly guided through the design process.

CONTENTS

LIFEBOAT STATION DEVELOPMENT PROJECT 1
ENGINEERING DESIGN REPORT 1
PROJECT OVERVIEW 4
    THE PROJECT BRIEF 4
    THE SITE 4
    DESIGN SPECIFICATIONS 5
    IMPOSED LOADS 6
DESIGN PRINCIPLES 7
    FUNCTION 7
    THE SITE 7
    STANDARDISATION OF COMPONENTS 8
    ARCHITECTURAL PRINCIPLES AND AESTHETICS 9
    SUSTAINABILITY 9
    HEALTH AND SAFETY 11
CONCEPT DEVELOPMENT 13
    DESIGN PRECEDENTS 13
    CONCEPT EXPERIMENTATION 14
    DESIGN INSPIRATION 16
    CONCEPT DESIGN 17
DESIGN DEVELOPMENT 19
    THE DESIGN PROCESS 19
    FRAMING PLANS 19
    SUBSTRUCTURE 25
    BRACING 27
    LATERAL STABILITY 30
    THE BRIDGE 33
    SLIPWAY 37
DESIGN PROPOSAL 39
    THE REVIT MODEL 39
    STRUCTURAL DETAILS 41
STRUCTURAL CALCULATIONS 44
APPENDIX 57

PROJECT OVERVIEW

THE PROJECT BRIEF

The Royal National Lifeboat Institute was established in 1824 and provided a 24-hour search-and-rescue service which operates in the coastal waters and significant inland waterways of the British Isles. The service currently operates from 236 lifeboat stations.

Ciocoiu+Milwid has been commissioned by RNLI to design a new lifeboat station on the Purbeck Heritage Coast, in Dorset. The client has provided an extensive brief, and this includes, apart from functional specifications, the requirements that:
1. The design be contemporary rather than traditional;
2. The design aesthetics reflect the nature of the local surroundings.

A cofferdam is to be constructed before the commencement of construction work on the lifeboat station. The client has commissioned an independent contractor to build the cofferdam, and this will therefore not form part of the construction programme proposed by Ciocoiu+Milwid.

THE SITE

The site is located on the Purbeck Heritage Coast, which is a nature conservation area and which is remote from urban development. The site is accessed via a narrow paved rural road.

The lifeboat station is to be constructed off-shore, near the rock cliffs. A submerged concrete slab remains from the earlier structure that has now been demolished. This slab will serve as foundations for the new design. A cofferdam is to be constructed before the commencement of construction, to expose the concrete slab.

Figure 1: Site photo

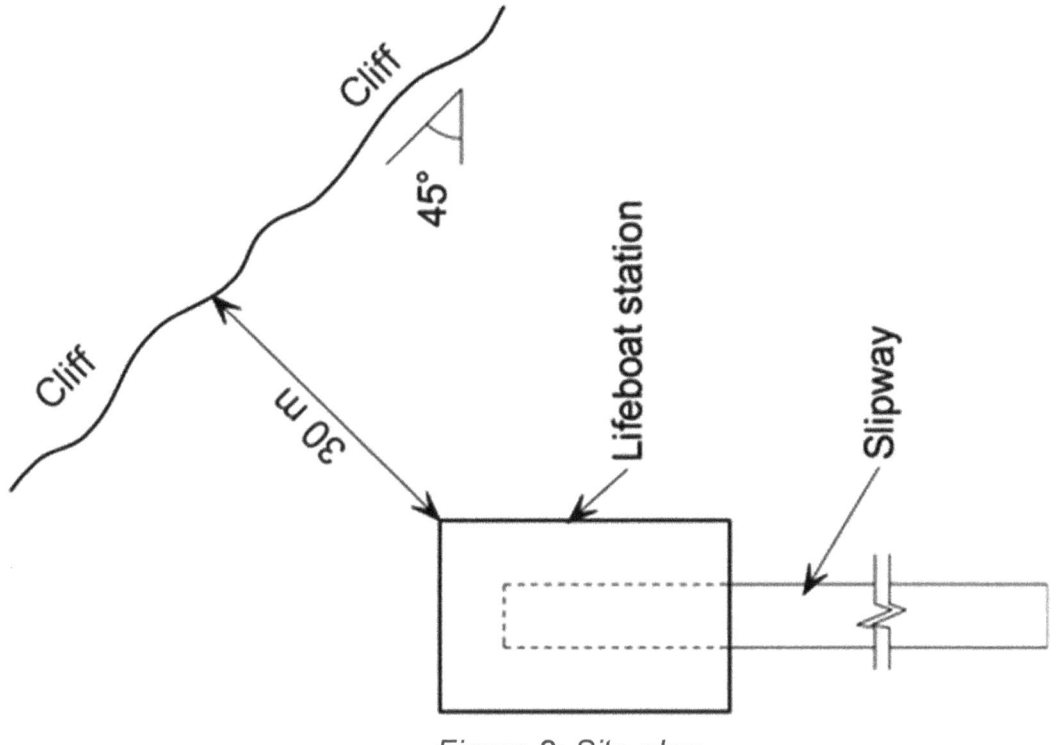

Figure 2: Site plan

DESIGN SPECIFICATIONS

	Description	Dimensions
ELEVATION ABOVE WATER LEVEL	Establish the distance between the water level and the structure.	2m above maximum high water.
CREW FACILITIES	Office space and a mess are to be provided.	<ul><li>Total usable floor space is at least 100m^2.</li><li>Access to upper floors is in addition to the minimum usable floor space of 100m^2.</li><li>Establish a headroom of 3m to the underside of any ceiling.</li></ul>
EXTERNAL WALKWAY	To surround the entire station.	<ul><li>Minimum 1.5m width</li><li>Balustrade 1.1m height</li></ul>
PEDESTRIAN ACCESS	<ul><li>Access to the main structure is to be via a footbridge.</li></ul>	<ul><li>Shortest distance to shoreline is 30m.</li></ul>

	• The footbridge is to be horizontal.	• The shoreline is at 45⁰ to the long axis of submerged foundations.
ACCESS FOR BOAT	Access for the lifeboat is to be at the ocean-end of the main structure.	Opening of 6m x 6m to be provided.
BOAT ENVELOPE	The boat envelope will accommodate the lifeboat and will allow access for maintenance.	• Length: 21m • Width: 6m • Height allowance: 6m
ACCESS PLATFORMS	A high-level access platform will be provided to each side of the lifeboat.	• Length: 21m • Width: 1.5m • Height above floor: 4m
CRANE	Provision is to be made for an overhead travelling crane with a lift capacity of 2 tonnes.	Crane must be able to lift anywhere in the 21 x 6m boat envelope.
LOCATION OF WINCH	Position the winch at the shore end of boat envelope.	Line of action of winch: 1m above floor level.
STRUCTURAL FRAMEWORK	Steel, following the requirements of BS 5950-1:2000 or Eurocode 3: Part 1.1: 1992.	

IMPOSED LOADS

WAVE FORCES	• 5kN/m^2 applied horizontally. • Due to the dynamic and impact effects of wave forces, a further factor of 1.25 should be applied to the standard combination load factor. • Wave forces should be taken to act on all submerged elements in the same direction at any one time. Wave forces can operate in any order.
WIND	• The dynamic pressure of 1.5 kN/m^3 • To be determined from BS6399-2
SNOW LOAD	0.6kN/m^2 on the roof and on any other area on which snow might fall.
FLOOR LOAD	7.5 kN/m^2 over the main concrete floor slab.
LOAD OF OFFICE AND MESS	2.5 kN/m^2 on any other story + 5% variable load
LIFEBOAT	Imposed in addition to loads listed above + 2% variable load

DESIGN PRINCIPLES

FUNCTION

The design process was pragmatic, driven by a series of design principles and, in a sense, resulted in the structure almost designing itself. The principles that guided the design are outlined over the following pages.

The fundamental principle underlying the design process was "function". The primary function of the proposed structure is to serve RNLI rescue operations. Thus, the starting point for the design was the boat envelope itself. The weight and positioning of the lifeboat, in turn, drove the creation of the substructure.

The requirement for a winch (for hauling in the boat) dictated the dimensions of the functional areas.

Entrance to the boat envelope was necessary to the south end of the structure. Although the public will be admitted to the building, the system has an operational function and is designed primarily to facilitate that function.

THE SITE

The site imposes three unique constraints on the design process:

LIMITED ROAD ACCESS

The site is in a rural location, with limited road access. Consideration must therefore be given to how equipment and materials will be transported to the site. The Design Team considered, early in the process, that transport by a sea route was likely to be the best option.

NATURE CONSERVATION AREA

The site is in a nature conservation area and thus demands vigilance about sustainability and protection of the natural environment.

SPACE LIMITATIONS

The construction site is to be exposed by erecting a cofferdam. Space on the site will therefore be minimal, and the demand for storage of materials and equipment on-site must be reduced as far as possible. Consideration should, therefore, be given to having materials delivered as required rather than in advance. Secondly, care must be given to eliminating the need for an on-site crane. The Design Team considered that these factors point towards the use of a barge-mounted crane.

The smaller the structural components, the safer and more efficient would their handling on a remote site be. Furthermore, the use of smaller parts reduces manufacturing and transport costs. In consideration of this constraint, the design team endeavoured to limit the length of any one type of component to 5 metres.

STANDARDISATION OF COMPONENTS

SYMMETRY

The grid was designed with symmetry along the long axis. Balance provides several advantages:
1. Assessment of stability: if the design fails, then whatever adjustments are required for the one side can be duplicated for the other side.
2. Reduced number of different type of components: the components used on one side of the axis of symmetry are duplicated on the other.

STANDARDISATION OF COMPONENTS

The use of standardised components is more efficient and more cost-effective than the use of custom-made parts. Standardised components offer a series of advantages:
1. They can be ordered "off the shelf" and are therefore readily available, with the benefit of reducing delays;
2. They are more readily packed and transported;
3. Standardised construction methods can be used, which allows for quicker construction time;
4. Using standardised components allows for repetitive processes which in turn allows for greater efficiency in construction;
5. The wastage that results from manufacturing custom-designed components is eliminated if standardised components are used.

The proposed design incorporates a limited number of lengths of beam and columns, and repetitive patterns are used in the framing plans. At a reasonably advanced stage in the design process, the Design Team realised that they had missed several opportunities for designing components of standard lengths into the framing plan. For example, the columns in the substructure are designed at 4.4 m in size. It might have been useful to round this up to a length of 4.5 or even 5m.

DIMENSIONS

Limiting the dimensions of components has several advantages:
1. Parts that are of a size that they can be readily transported allow for greater efficiency and are therefore more cost-effective.
2. They can be handled more quickly if there is consistency in size;
3. Small components and light can be carried and put into place by human operatives (rather than depending on lifting equipment and cranes)

MODULAR COMPONENTS

Modular components, such as for cladding and concrete floors, reduce the need for workspace on-site. The use of modular concrete block flooring means that concrete mixers are not required for casting concrete on site.

REPETITIVE PROCESSES

Using standardised components and reducing the number of types of features allows for repetitive construction processes. Operatives are more efficient when doing repetitive processes, and this has a positive impact on construction time and costs.

ARCHITECTURAL PRINCIPLES and AESTHETICS

Modernist Design principles and the notion of being honest to materials allows for structural elements to be exposed and to be made a feature. Reference to this principle drove several design decisions, such as the following:
1. Bracing becomes an aesthetic feature of the structure;
2. A glass curtain wall across the entire face of the cliff-end of the main structure is placed to showcase the engineered designs within;
3. Mechanical and electrical elements will remain exposed and will be enhanced with coloured paint;
4. Internal structural elements (such as columns) are left exposed.

SUSTAINABILITY

BREEAM (Building Research Establishment's Environmental Assessment Method) is a sustainability rating scheme for the built environment.

For a new structure such as this lifeboat station, the designers would refer to the requirements of the *BREEAM New Construction 2014 Non-domestic Buildings (SD5076:41)*, which describes the environmental performance standard against which new, non-domestic buildings in the UK can achieve a BREEAM New Construction rating and how the facilities are assessed.

Assessments are based on a scoring system and are carried out against nine criteria:
1. Management
2. Health and wellbeing
3. Energy
4. Transport
5. Water
6. Materials
7. Waste
8. Land-use and ecology
9. Pollution

These criteria and their subheadings take account of the entire lifespan of a project, from conception to demolition. Taking these criteria into account during the design phase is a practical approach since, when a local authority returns a decision on a planning application, the decision is usually subject to a list of conditions and one of these conditions is often a requirement to meet a specific BREEAM assessment rating.

Since the process of completing the detail design for this project and the subsequent construction phase is necessarily a collaborative exercise – between Ciocoiu+Milwid, contractors, M&E engineers, consultants and the project manager – for a sustainability strategy to be successful, all parties involved would have to commit to the plan.

During the design phase, each stage in the lifespan of a structure should be designed with sustainability considerations in mind. The members of our Design Team do not profess to be experts in the matter of sustainability and, in progressing further with this design, would recommend that an expert consultant in this field to oversee the remainder of the design process, and that a site sustainability manager is appointed to ensure that the construction site is managed in an environmentally efficient manner.

In the paragraphs below, reference is made to some of the sustainability considerations.

THE DESIGN PHASE

From a sustainability point of view, the design should incorporate construction materials with a low level of environmental impact. For a coastal site, where the structure will inevitably be subjected periodically to aggressive weather conditions and the effects of landing and launching a lifeboat, there is little option but to use steel for the supporting structure.

Steel has a relatively high level of embodied energy. However, this can be offset by using other materials. Additionally, as an element of the sustainability plan, the principal contractor will be required to source all building materials in compliance with a sustainable procurement plan.

THE CONSTRUCTION PHASE

The responsibility of minimising the impact of the new lifeboat station on the environment during the construction phase is mostly on the main contractor. As part of the sustainability strategy, the procurement process would involve appointing a contractor who complies with a local or national Considerate Construction Scheme. Ideally, also, a site sustainability manager would be assigned to ensure that the construction site is managed in an environmentally efficient manner.

From the design point of view, a significant factor in planning for the impact of the construction phase to have as low an impact on the environment as possible is to design for the construction phase to be completed as quickly as possible. The shorter the time between commencement of construction and handover, the less damage there is likely to be to the environment and the lower the level of disturbance to local flora and fauna. As outlined in earlier sections, the use of pre-fabricated and standardised components increases efficiency during the construction process and would be a significant factor in ensuring that construction is completed promptly. Additionally, delivery of machinery, equipment, and materials via a sea rather than the land route, would have less impact on the natural land habitat – although care would have to be taken not to cause undue disturbance to marine life.

THE OPERATIONAL LIFE-CYCLE

The structure will, over its life-cycle, be subjected to extreme weather conditions and heavy operational use. The impact on the environment of maintenance is reduced if materials that have a good life span and a low tendency towards degradation are used. The use of modular components, such as for cladding and roofing, means that individual parts can be replaced during maintenance, which has less impact on the environment than replacement of the entire feature (such as roof covering).

DEMOLITION

The design should allow for the structure to be demolished with minimal disruption to the environment. The use of recyclable materials reduces the amount of waste generated during demolition. As a construction material, its heavy carbon load mitigates against steel. But steel comes into its own at the demolition phase, since it is readily recycled. However, in the case of a structure like a lifeboat station, re-purposing offers an even better alternative than demolition – as can be seen in the image below.

Figure 3: Purbeck heritage coast lifeboat station

HEALTH AND SAFETY

CDM Regulations, Health and Safety Regulations, and Building Regulations (amongst others) must be considered during the design process, to ensure compliance during construction and during operation of the structure during its useful life.

The Design Team admits to not being experts in this field and, for future stages of the development of this design, intend to appoint an expert advisor.

The nature of the construction site brings with it a series of extraordinary risks relating to construction and the future operation of the structure, that must be considered during the planning phase.

DURING CONSTRUCTION

The construction site will be exposed by the construction of a cofferdam around it. The use of a cofferdam brings with it the risks of a breach. Emergency systems should be put into place to prepare for the eventuality of failure of the dam.

The delivery of materials and equipment by barge must be carefully planned.

The risks associated with aggressive coastal weather conditions must be planned. In terms of the principles outlined in earlier sections, delivery of materials and equipment will be by sea route. Additionally, the site will be served by a barge-mounted crane. This strategy eliminates the need for a crane to take up space on the building site itself, but it does bring with it the risks associated with operating a crane from an unstable base: a barge being battered by winds and high seas.

Working on a site that is surrounded by water brings with it the higher-than-usual demand for management of the risks of using with electricity Services should be positioned to protect against the risk of contact with water and the risk of being penetrated by humidity. The risk of slippages associated with wet surfaces must be mitigated.

DURING THE OPERATION LIFE OF THE STRUCTURE

While lifeboat station personnel have been trained to manage the risks associated with working at sea and in a marine environment, members of the public who visit the lifeboat station may not have. The risk of personnel and visitors falling into the sea from the deck of the lifeboat station must be mitigated. Balustrades alone are not sufficient – and must, also, be prepared to protect children from slipping between railings.

The risk of slips and floors must be mitigated. The floor surface should be prepared to have slip-free characters. This is true not only for the external floor surface (such as on balconies) but internally as well since, in a coastal location, the risk of interior floors becoming wet is high. Horizontal surfaces of exterior parts of the structure (such as walkways and the bridge) would best be designed with perforations to allow drainage.

MAINTENANCE

About care, three key factors (amongst others) must be considered during the design process.

THE RISK OF CORROSION

Parts of the substructure will be permanently underwater, while that above water-level will be subjected to ocean spray. The design should incorporate materials that have been chemically protected against corrosion, to limit the requirements for maintenance.

WEATHER CONDITIONS

The structure will be subjected to weather conditions that at times will be aggressive. The design should incorporate components that can withstand high winds and heavy seas, to minimise the requirements for maintenance. Careful attention should be paid to designing appropriate fixings for the cladding and roofing.

ACCESS

The remote location and the limited road access mean that routine maintenance once the station is in operation should not require the use of heavy machinery or replacement of heavy components. The glass curtain wall at the shore end of the main structure would, for example, ideally be constructed from self-cleaning glass (such as that manufactured by Pilkington).

CONCEPT DEVELOPMENT

DESIGN PRECEDENTS

Before starting work on the concept design, the Design Team looked at several lifeboat stations around the country, to get ideas about form, programming and aesthetics. A selection of stations is shown in the images below.

TENBY LIFEBOAT STATION

Pembrokeshire

BEMBRIDGE LIFEBOAT STATION

Isle of Wight

ALDEBURGH LIFEBOAT STATION

Suffolk

WHITSTABLE LIFEBOAT STATION

Kent

LIZARD LIFEBOAT STATION

Cornwall

St. ABBS LIFEBOAT STATION

Berwickshire

CONCEPT EXPERIMENTATION

The concept design was generated through a process of exploration of options, such as for roof shape, programming of spaces and aesthetics. The process began with hand sketching. Later, some of the ideas were transferred into digital models using AutoCAD or Revit, as shown in the table below.

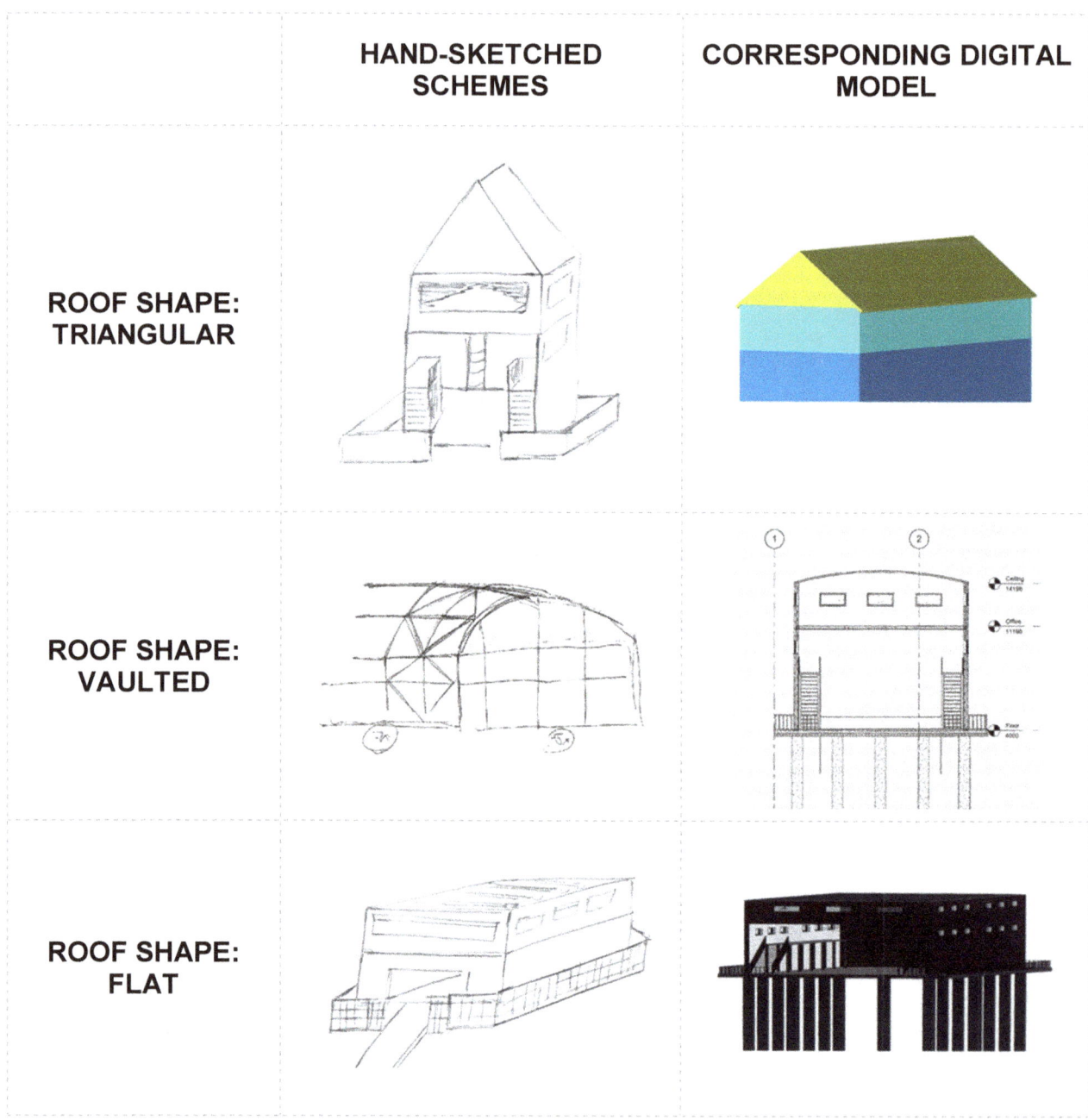

HELICOPTER LANDING PAD		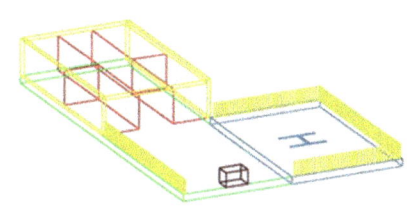
STAIRS EXPOSED TO ELEMENTS Stairs could be placed outside, to make the most of the available space.	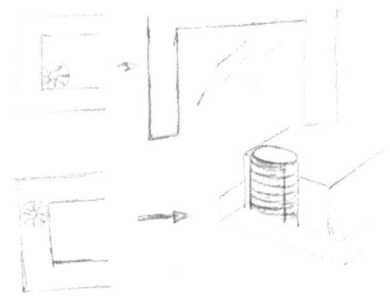	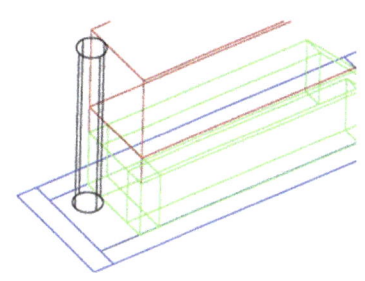
A TALL THIN STRUCTURE WITH MULTIPLE STORIES: Boat on the lowest storey, other facilities (office, mess) on a higher level.		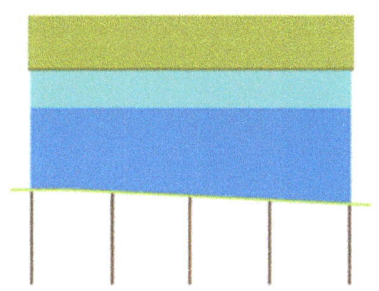
A LONG NARROW STRUCTURE	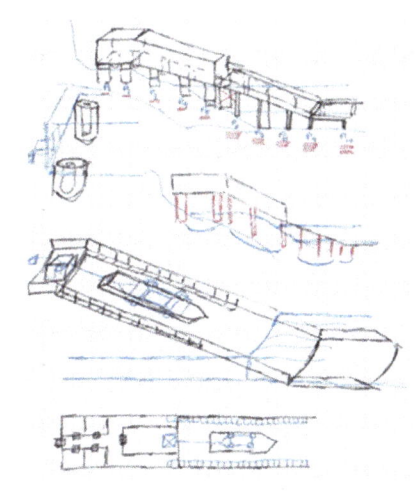	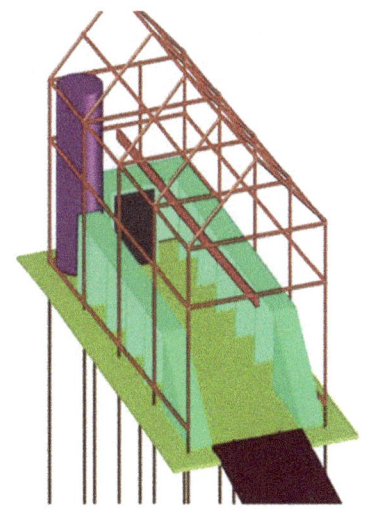

A LOW FLAT STRUCTURE

All facilities (the boat, offices, mess and entrance) are based on a single level.

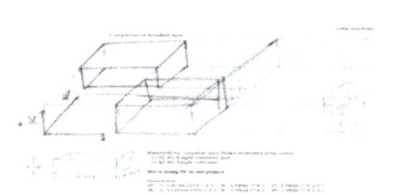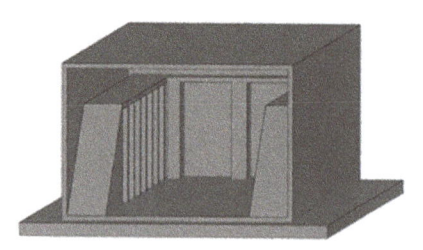

DESIGN INSPIRATION

The lifeboat station is to be constructed in a bay, surrounded by rocky cliffs, on the Purbeck Heritage Coast, and the client's brief specified that the nature of the local surroundings is reflected in the design aesthetics.

The Design Team felt that the local surroundings provided plenty of inspiration for the design. The marine location, the rocky cliffs, and the maritime context were the primary drivers for the concept design, as illustrated by the images below.

INSPIRATION FROM THE OCEAN	INSPIRATION FROM THE TERRAIN	MARITIME INSPIRATION

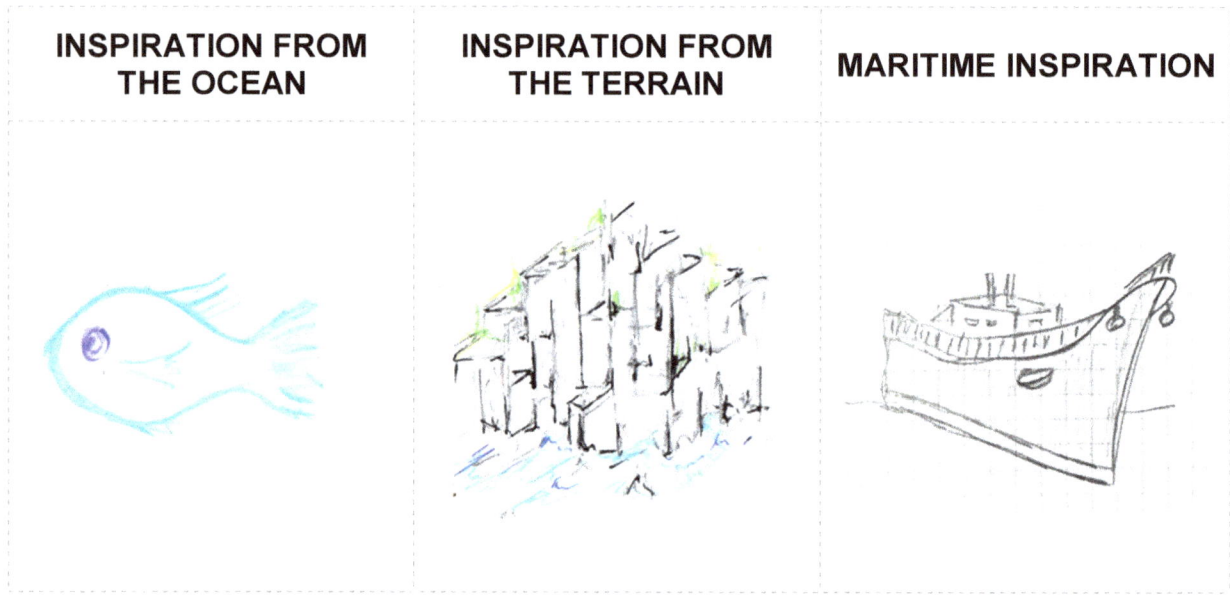

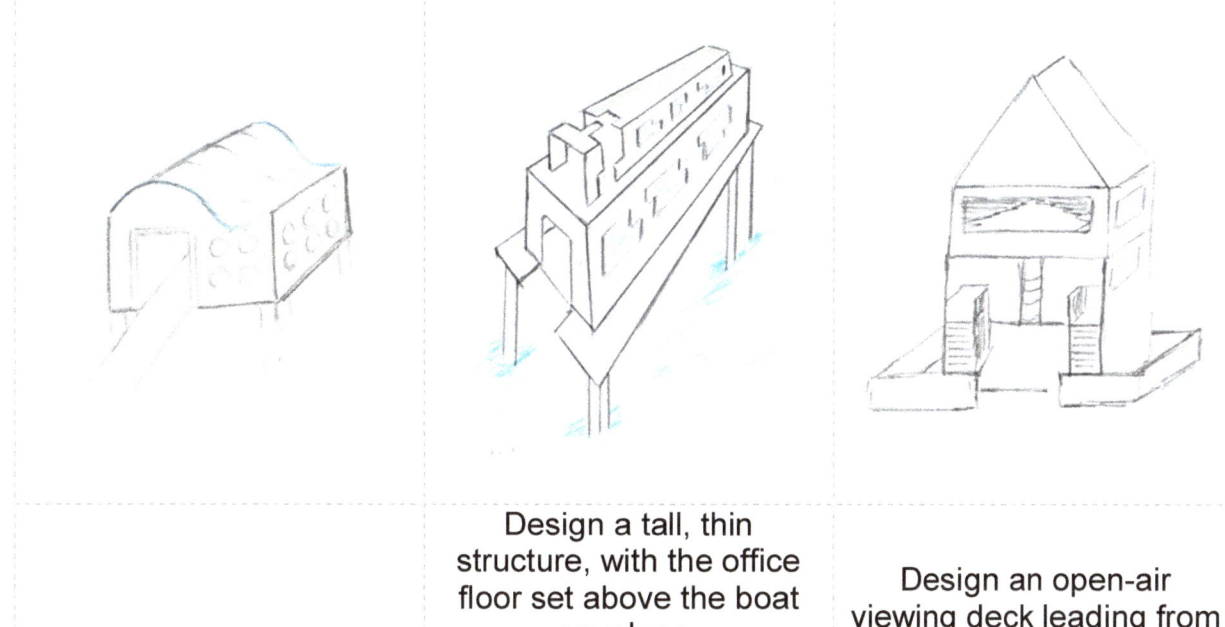

Design a vaulted roof.

Design a tall, thin structure, with the office floor set above the boat envelope.

Establish an open-air walkway surrounding the office space.

Design an open-air viewing deck leading from the office space at the south end of the main structure.

CONCEPT DESIGN

The concept design for the main structure combines elements of the design schemes shown on the previous pages.

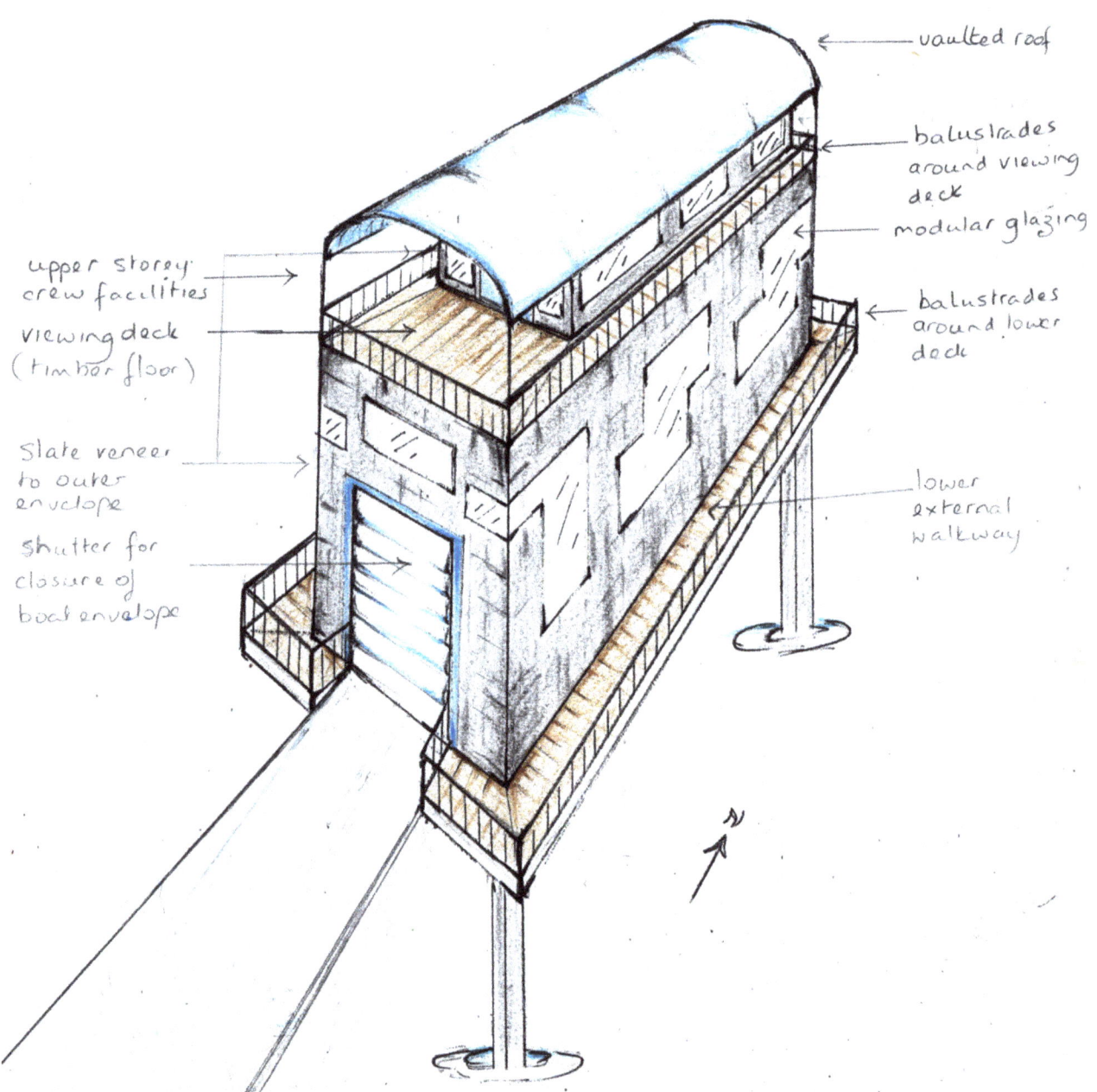

Figure 4: Concept design

DESIGN DEVELOPMENT

THE DESIGN PROCESS

Development of the design was driven by elements of the design brief, together with the design principles outlined in an earlier section.

The two initial objectives of the process were:
1. To establish a provisional grid and levels, on which to base development of a Revit model of the structure;
2. To establish a framing plan on which to base the Revit model.

The grid and levels for the Revit model were generated through a pragmatic process, as illustrated in the Book 2 of this report, whereby the design specifications were applied following the design principles outlined in the earlier section. The framing plans were generated through a similar process.

These processes are outlined in Book 2 of this report, in the following order:
- THE GRID
- THE LEVELS
- FRAMING: PLANS
- FRAMING: ELEVATIONS
- FRAMING: SUBSTRUCTURE
- BRACING

FRAMING PLANS

The process whereby the framing plans were generated is outlined in detail Book 2 of this report. The diagrams below show the elevations that were the outcome of this process.

ELEVATIONS: ORIGINAL VERSION

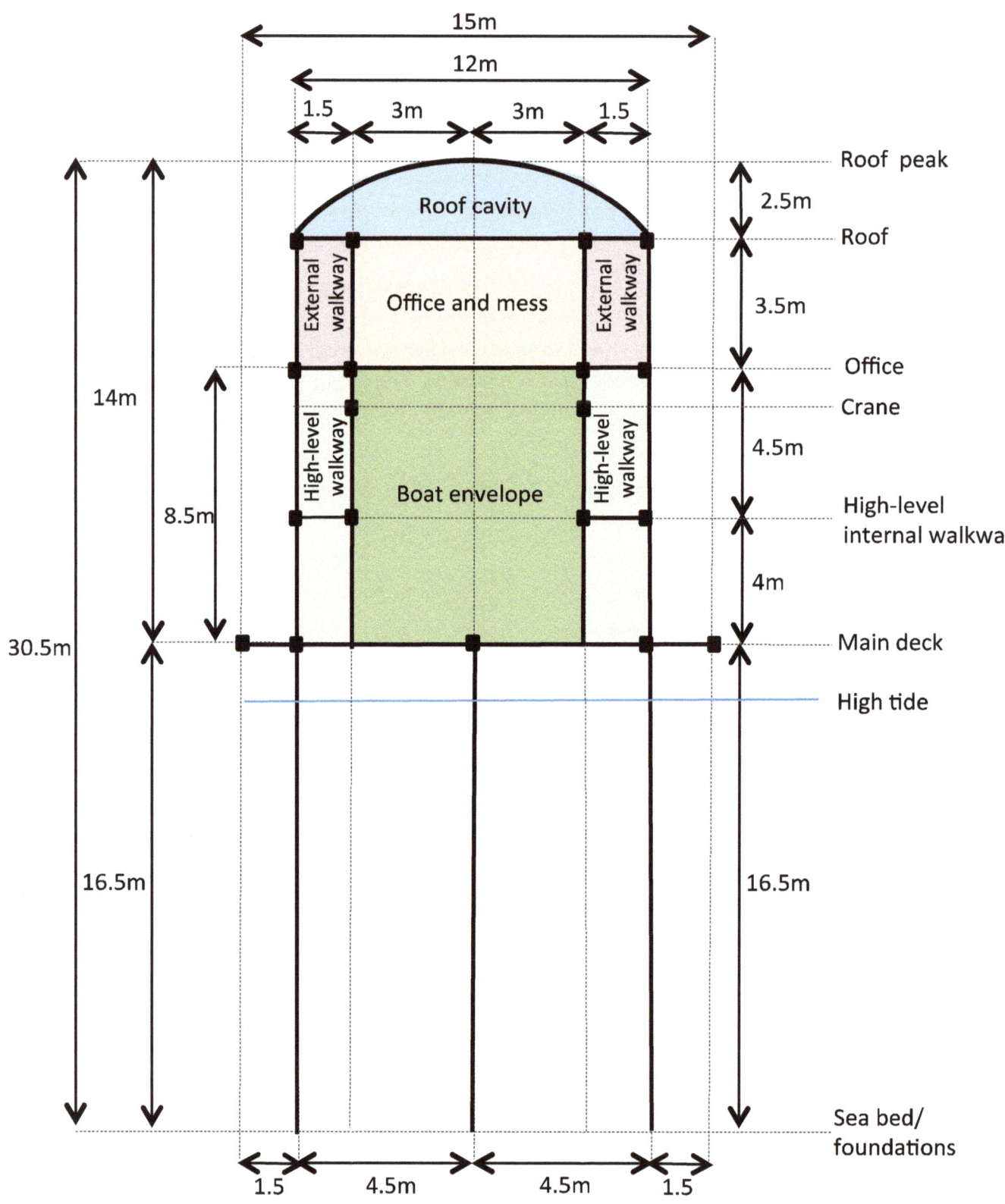

Figure 5: South elevation

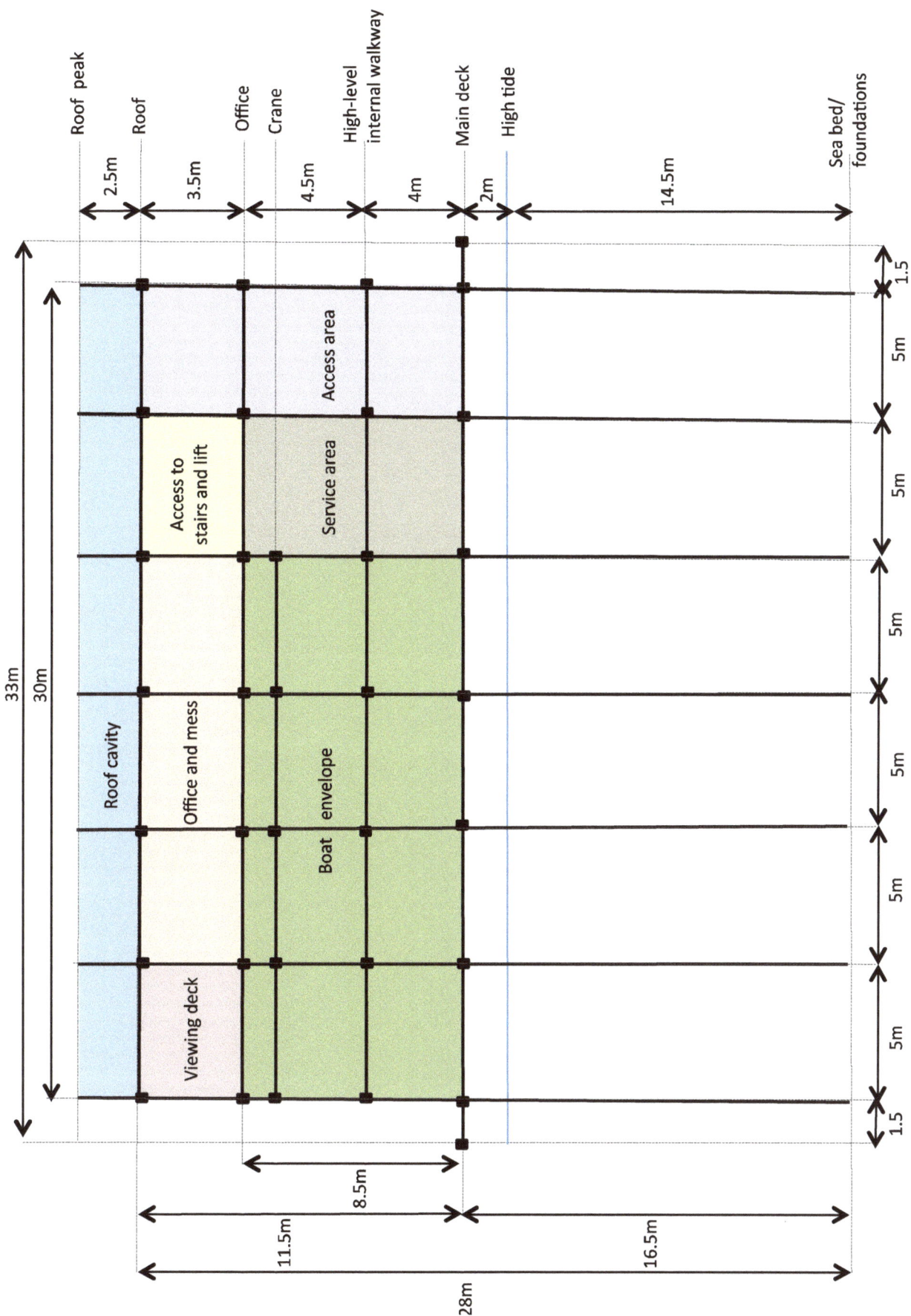

Figure 6: East elevation

ELEVATIONS: REVISED VERSION

The original version of the north-south elevation and framing plan, as shown in the diagrams on the previous page, assumed that a ceiling would be installed at the office level.

In later design development, a decision was made to eliminate the ceiling to the offices and instead to have a vaulted roof cavity to the office area. The framing plan was adjusted accordingly, as shown in the diagrams below.

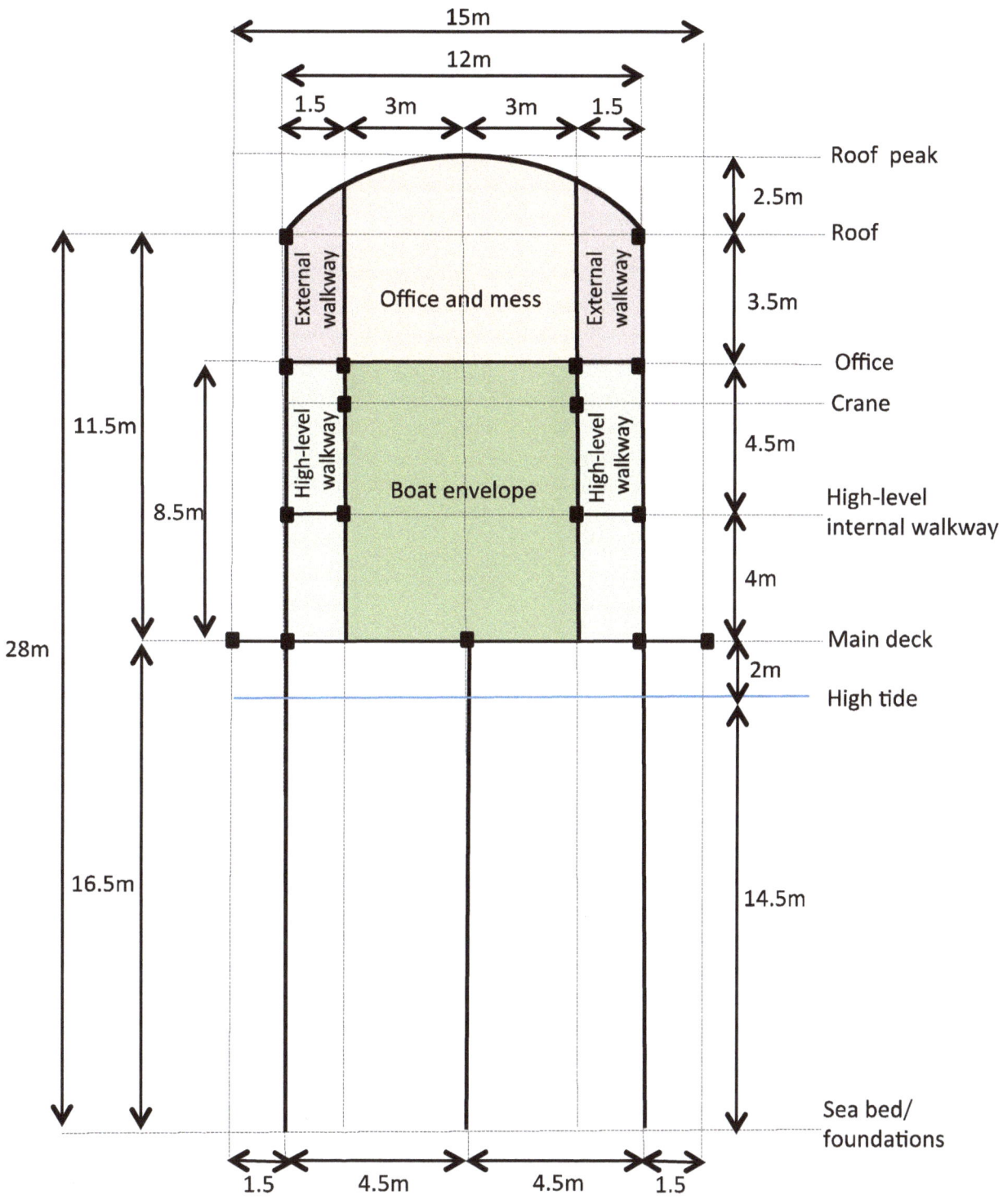

Figure 7: South elevation

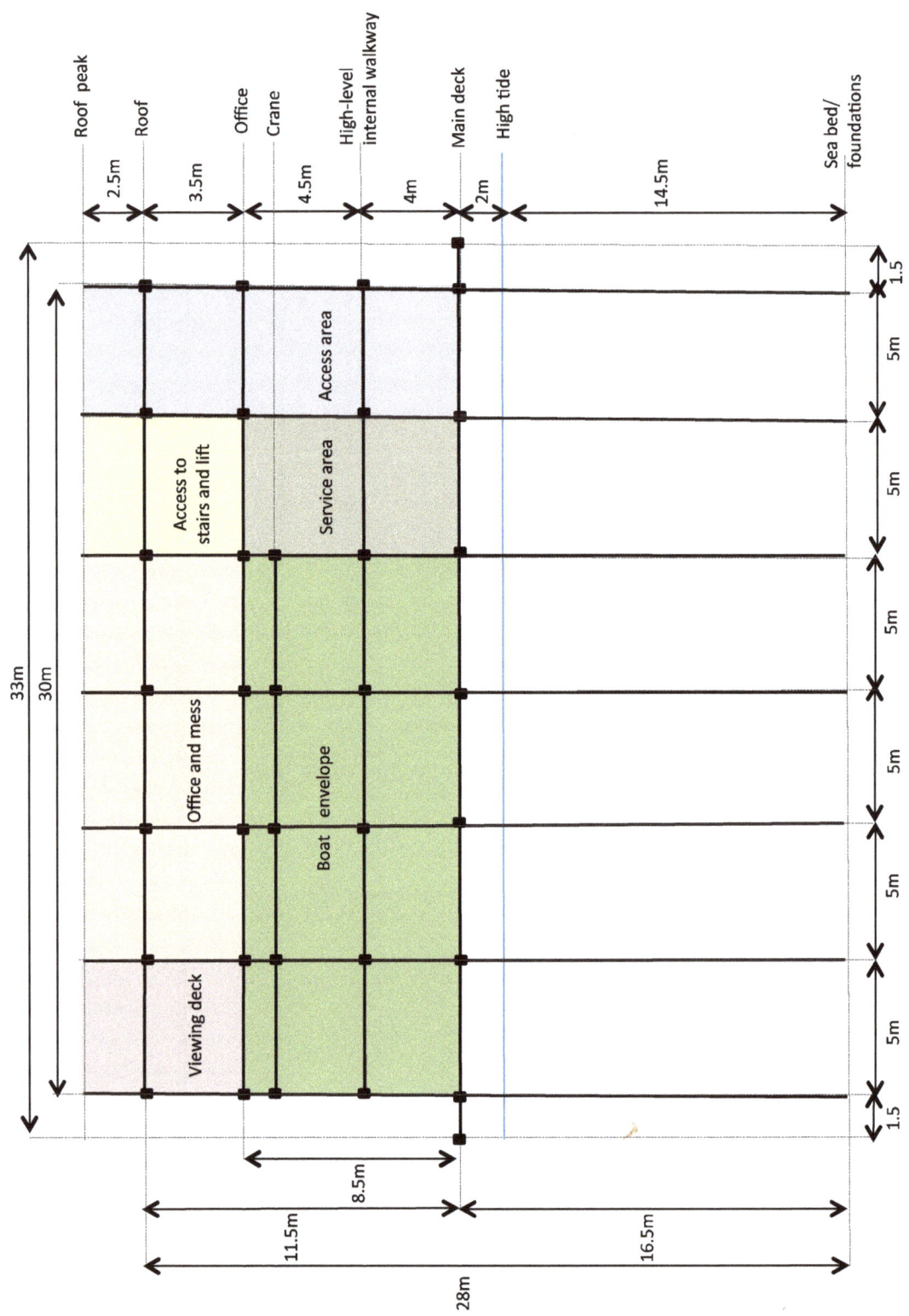

Figure 8: East elevation

SUBSTRUCTURE

The height of the main deck above the foundations is considerable: 16.5m. It was decided, for preliminary planning purposes, to subdivide this height into evenly distributed 4m components, as shown in the diagrams below. In the Revit model, after adjustments were made for various factors, the length of these components was adjusted to 4.25m.

Beams are placed to connect the columns, as shown in maroon.

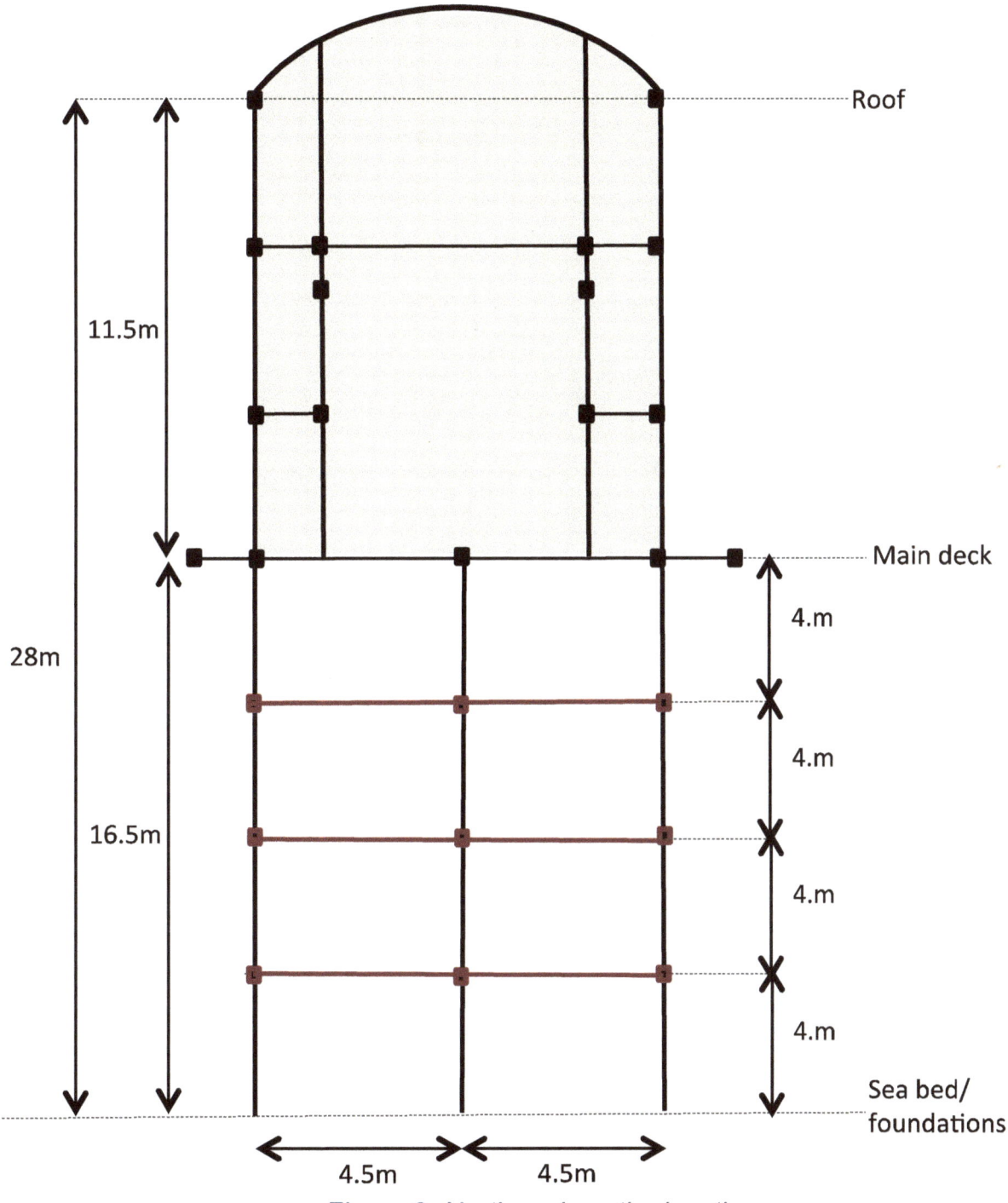

Figure 9: North and south elevation

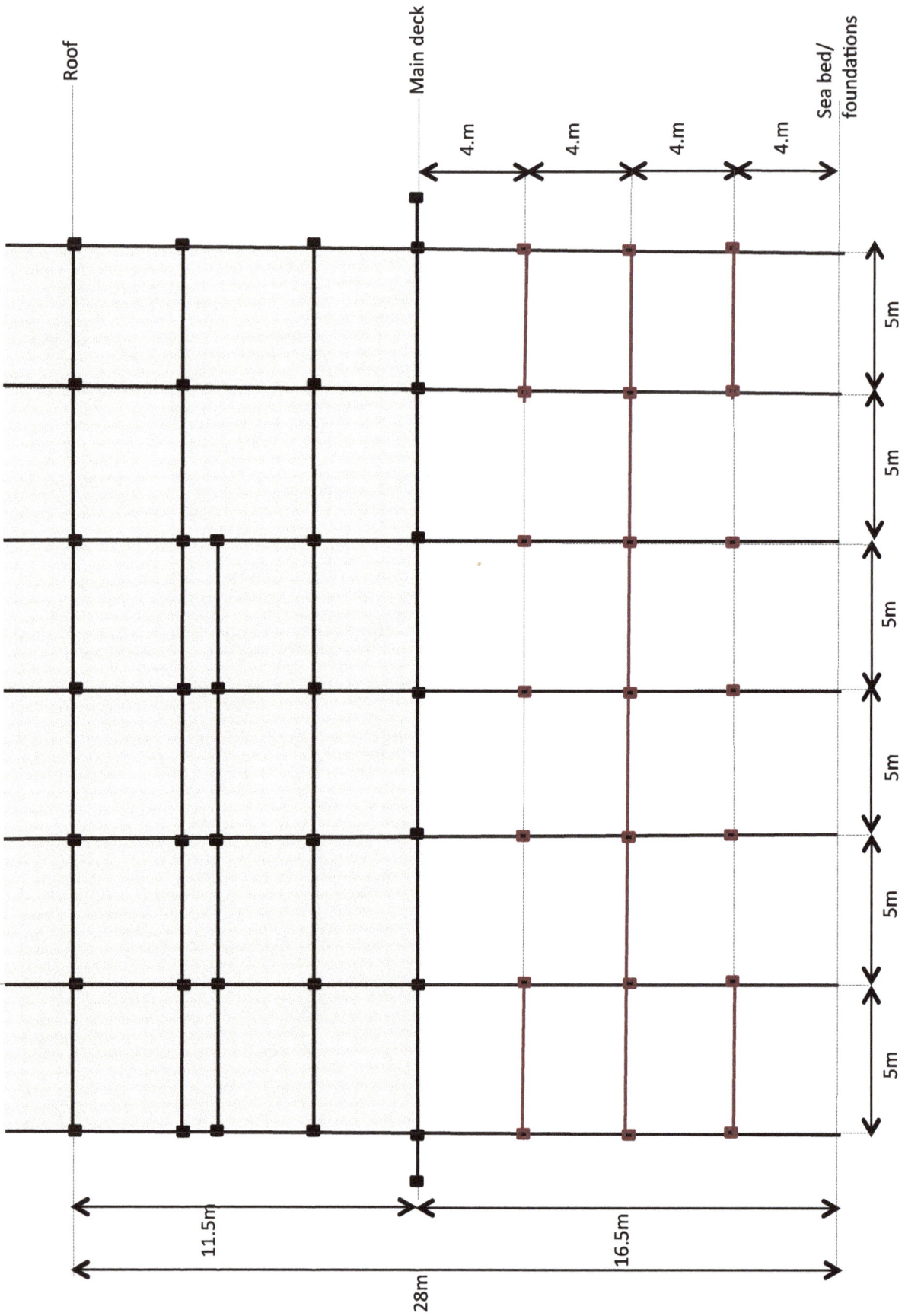

Figure 10: East elevation

BRACING

A design decision was made to employ bracing (shown in orange in the diagrams below) to provide stability in the north-south direction and portal frame action to provide stability in the west-east order.

In the west and east elevations, bracing is provided in each of the end bays. Bracing is provided in more than one bay, to protect against the risk of failure of the bracing in the other bay.

In the north and west elevations, stability in the superstructure is to be provided by portal frame action. However, in consideration of the forces to be exerted on the substructure by wave action, bracing is placed at the substructure level of this elevation too, as shown in the diagram below. The issue of stability is discussed in greater detail in the following section.

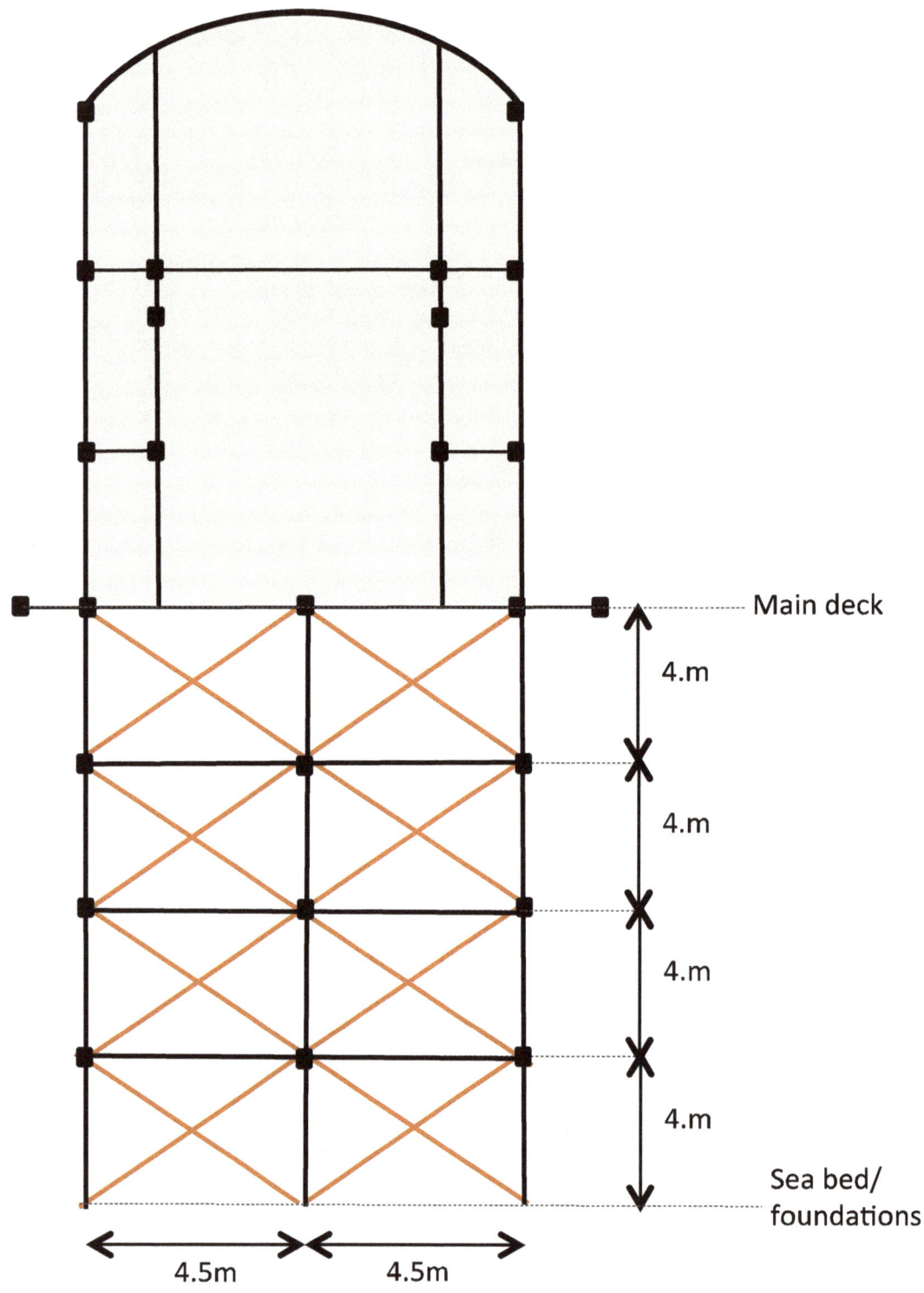

Figure 11: North and south elevation

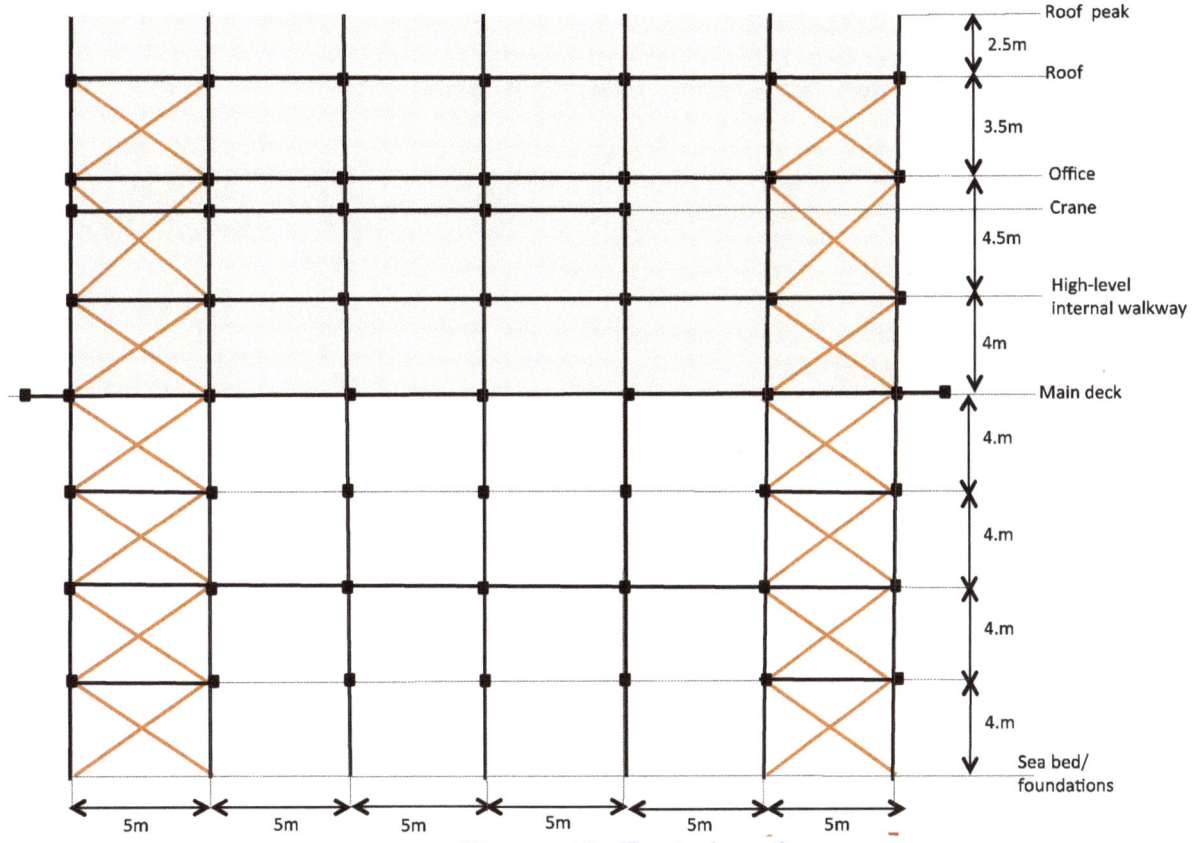

Figure 12: East elevation

LATERAL STABILITY

LATERAL STABILITY IN THE NORTH-SOUTH DIRECTION

<u>The Superstructure</u>
Since the structure is in a coastal location, it is likely to be subjected to extreme weather conditions (both wind and wave actions). For this reason, extra measures are taken in terms of bracing:
1. Two bays of bracing are provided on each façade to protect against the risk of one bay failing.
2. Cross-bracing rather than simple diagonal bracing is provided

<u>The Substructure</u>
As with the superstructure, bracing is provided in the outer bay at either end of the substructure. Additional stability is achieved by bracing the column of each substructure to the adjacent column(s) at mid-height.

LATERAL STABILITY IN THE WEST-EAST DIRECTION

<u>The Substructure</u>
Since the substructure is likely to be subjected to considerable wave forces, cross-bracing is provided, as shown in the diagram on the previous page. This bracing pattern is repeated at each of the seven sets of columns.

<u>The Superstructure</u>
Resistance to wind and wave forces from the west and east is achieved by portal frame action in the superstructure and bracing in the substructure. The choice of a portal frame for the superstructure was based on the requirement to have an obstructed 6m opening at the south end of the structure and a free 6m-width boat envelope.

At a relatively advanced stage of the design process, however, our design advisors observed that, because the structure has a height of 19.6m above foundation level and yet is very slim, portal action in the superstructure might not provide sufficient stability. As a result, the Design Team was advised to incorporate bracing into the north and south elevations.

However, by this stage, the Design Team had by considerable progress in the building of the Revit model. So, a decision was made to reassess the design as it stood. In the discussion, one of our design advisors pointed out that the system, as it stood, represented a battened portal frame (as illustrated on the following page) and that a battening mechanism serves to strengthen the portal frame action.

Furthermore, there is not only one level of the portal frame, but two: a lower level frame to accommodate the lifeboat as well as one at office level, as shown in the diagrams on the next page.

As a result, it was decided not to make any further adjustments about stability in the west-east direction.

THE PORTAL FRAMES

The two levels of portal frame referred to in the previous paragraph are illustrated in the diagrams below. The portal frames and rigid joints are shown in green. All other joints are pinned.

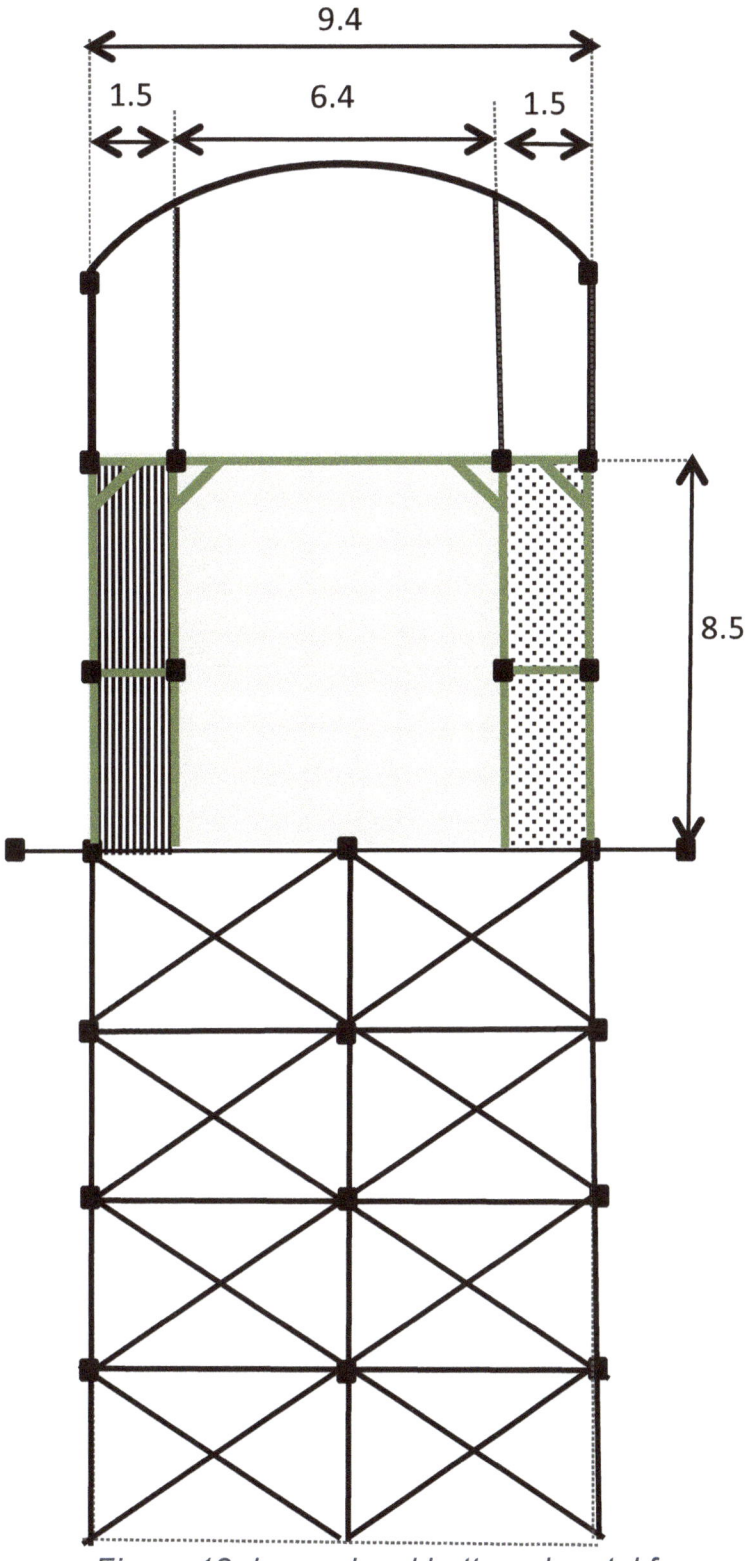

Figure 13: Lower level battened portal frame

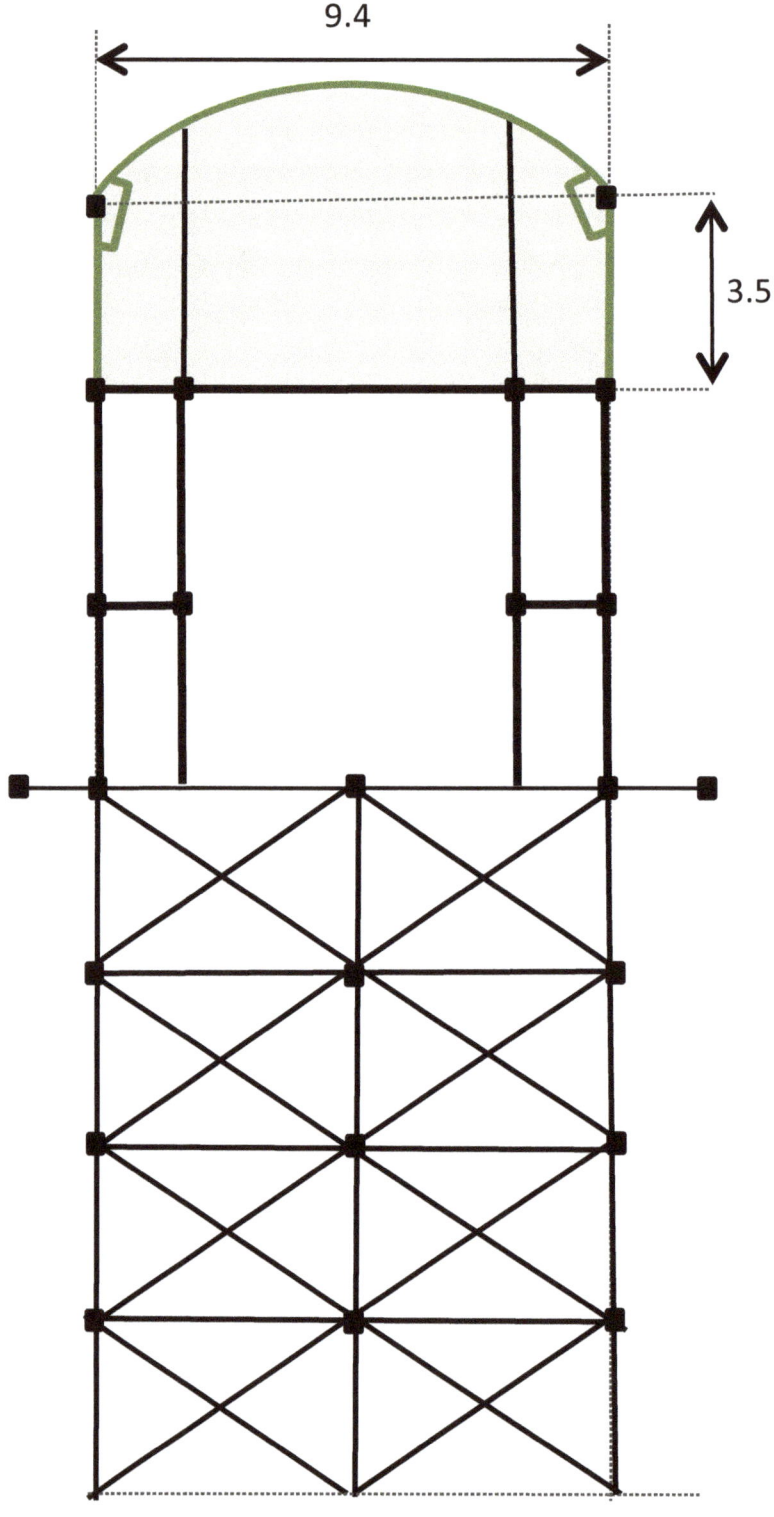

Figure 14: Upper-level portal frame (shown in green)

THE BATTENED PORTAL FRAME

The battened frame represents one portal frame lying within another, as illustrated on the following page. Each frame on its own is subject to sway. The battening mechanism, however, serves to stabilise the machine and provides resistance to the sway force imposed by the wind.

Bending moment diagrams have not been derived for these frames, but this would be a necessary process during further development of this design.

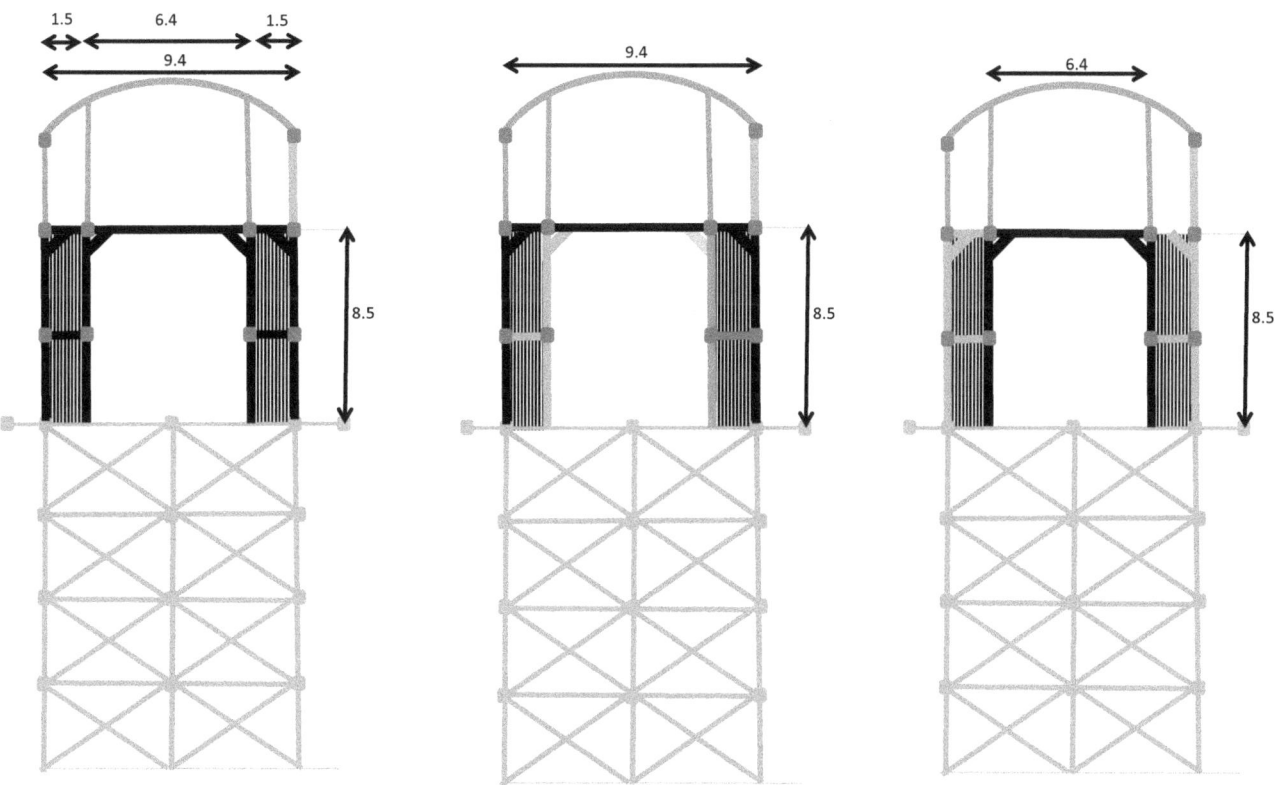

Figure 15: The battened portal frame and its elements

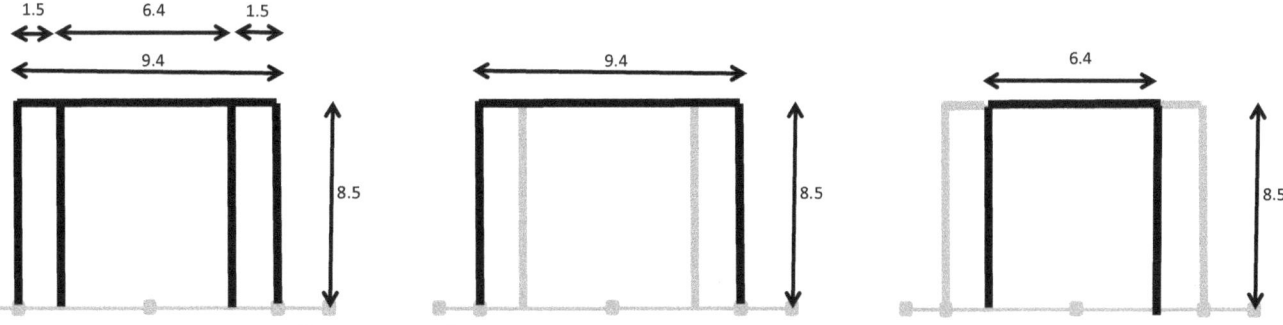

Figure 16: The battened portal frame and its elements

THE BRIDGE

A bridge is to connect the main structure to the cliffs and a pedestrian path that leads to the rural road.

The design for the pedestrian bridge is to be developed by a different consultancy practice, but our Design Team has produced an initial concept for the bridge, as shown in the diagrams below.

To minimise costs, construction time and environmental impact, the bridge must allow for the shortest route possible between the cliff and the structure.

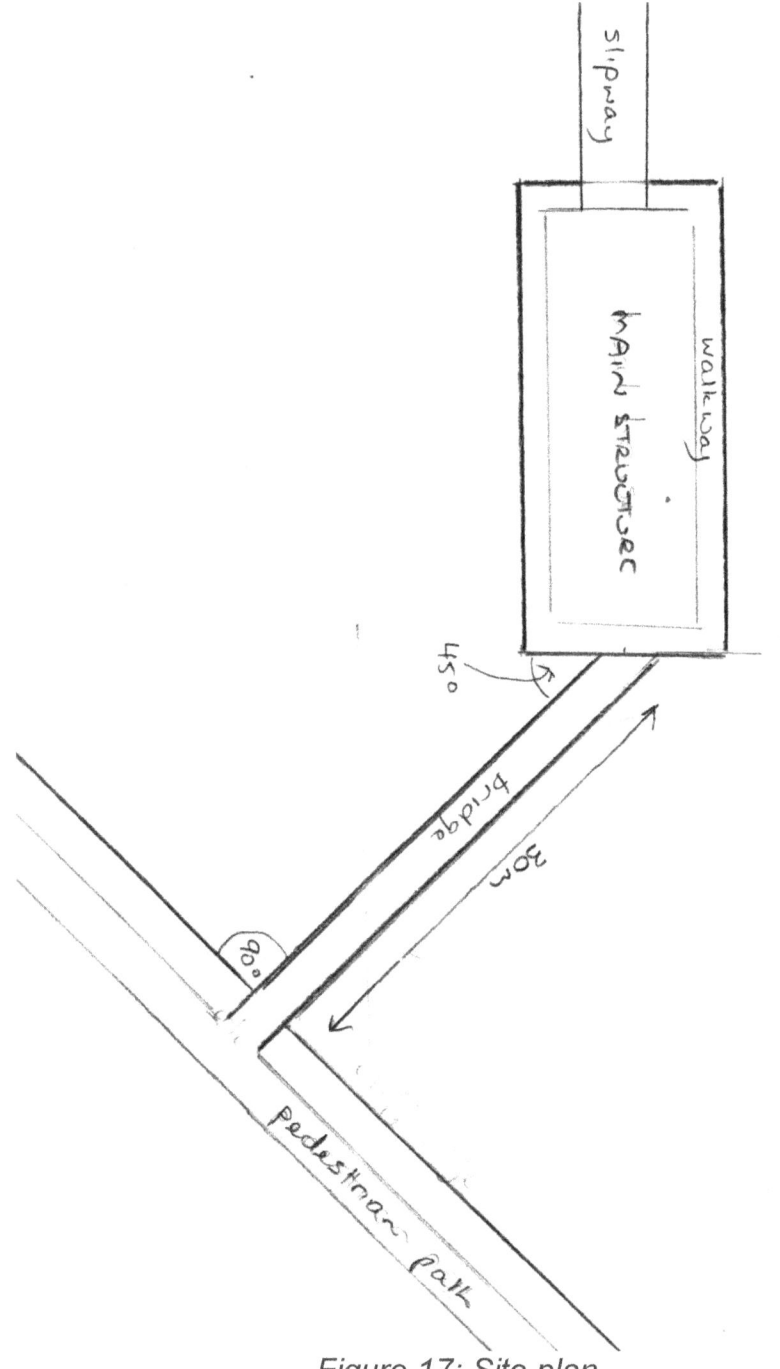

Figure 17: Site plan

The proposal illustrated overleaf was developed after considering several alternative structural options, which included the following:

1. Cable stay
2. Suspension
3. Simple beam
4. Lattice girders
5. Truss girders

The design team's preferred choice is a cable-stay bridge, with a cantilever structure, anchored at the shore end, as illustrated on the following page. The rationale for this preference is that:

1. The end adjacent to the main structure of the station does not require support and therefore imposes no load on the main construction of the station.
2. A cantilevered bridge can be designed to span the 30-metre distance between the shore and boathouse structure without intervening support columns.

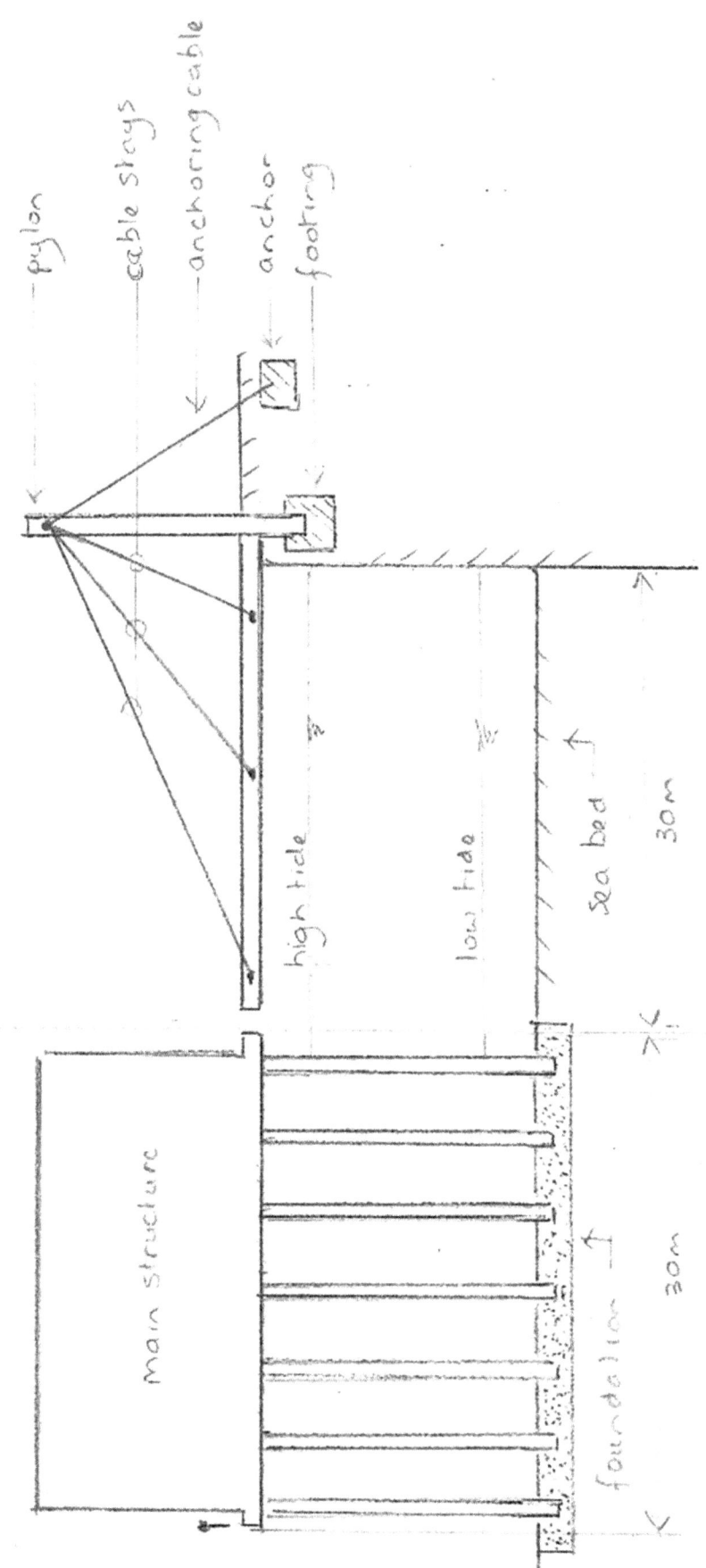

Figure 18: Concept proposal for a bridge

SLIPWAY

As in the case of the bridge, the design for the slipway has been outsourced by the client. However, our Design Team has produced an initial concept for the slipway, as outlined below.

SPECIFICATIONS

The cross-sectional specifications for the slipway are shown in the diagram to the right, as given in the client's brief.

These specifications are represented in the cross-sectional diagram to the right;

The gradient of the exterior section of the slipway is 11o, as per the design specifications given in the brief (as per the diagram below):

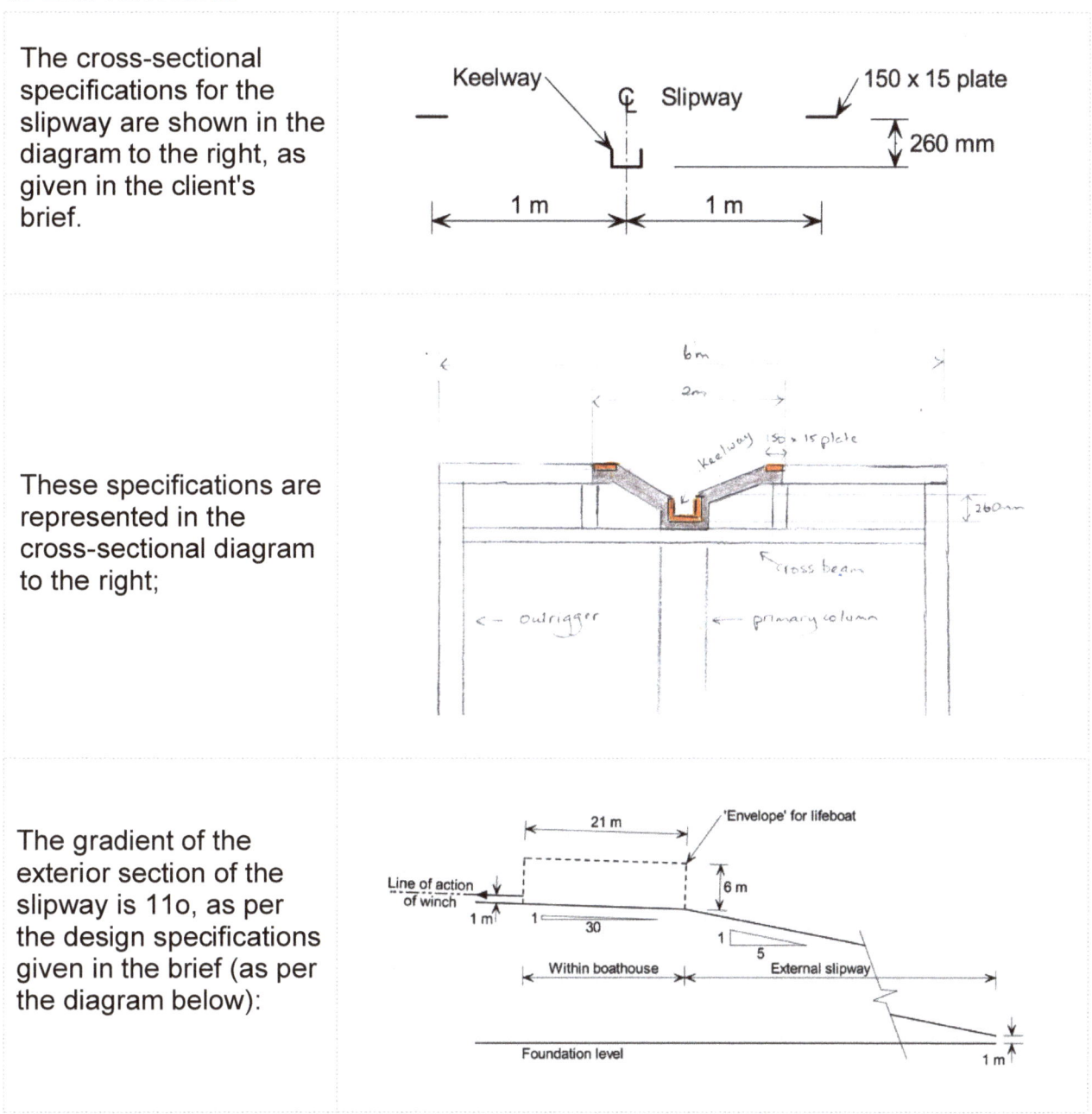

SUBSTRUCTURE

In section, a central steel structure of 2m width accommodates the mainframe and the 150mm plate. A walkway of 2m width runs alongside the central design on either side.

A primary column supports the central part of the load imposed by the lifeboat and outrigger columns support the extremes of the structure.

Stability is provided by bracing in both x and y directions at every second bay.

DIGITAL REPRESENTATION

The Design Team realised the proposal outlined above in a Revit model. Design sheets for the slipway are included in Book 2 of this report.

DESIGN PROPOSAL

THE REVIT MODEL

Once the draft framing plan had been generated, it was transferred to Revit, and a digital model was developed.

MODIFICATIONS

In generating the grid in Revit, a modification was necessary for that the draft framing scheme did not account for the thickness of columns. The specified width dimension of the boat envelope (6m) had been measured from mid-column to mid-column. Thus, in the Revit grid, this width was increased to 6.4m to accommodate the width of a column at each side. Similarly, the width of the external walkway was increased from 1.5 to 1.6m.

PROCESS

<u>Stage 1: Framing</u>
Framing for the main structure (apart from the arched roof beams) was systematically put in place as the first stage of development of the model.

<u>Stage 2: Roof beams</u>
A decision was made, under the guidance of our design advisors, to use arched castellated beams for the roof. Doing the necessary research and then drawing and designing this beam proved to be a considerable challenge. Our crucial reference source was Cellular Beams: a Design Guide, published by Macsteel Trading (www.macsteel.co.za/files/data_downloads/25/macsteel-trading-cellular-beams-design-guide.pdf).

<u>Stage 3: The outer envelope</u>
In terms of the design principle (as outlined in an earlier section), modular components were selected for cladding and the roof covering. Kingspan Insulated Roof and Wall Panels were chosen because they incorporate insulation, which eliminates a stage in the construction process (i.e. the installation of insulation). The manufacturer's product specifications are included in this report. The north elevation is glass curtain-walled in its entirety.

<u>Stage 4: Floors</u>
Modular components are once again incorporated into the design, with block-and-beam concrete floors are used (product specifications are shown in Book 2 of this report).

<u>Stage 5: Access</u>
In line with building regulations, a lift has been built into the model, to allow access step-free access to the office floor. The stairs allow access both to the office floor and to the high-level walkway on the east side of the boat envelope. Access to the high-level walkway on the west side of the boat envelope is provided using stairs from the winch

deck. Since the winch deck is an operational area and not open to the public, the west side walkway will be accessible to personnel only.

Stage 6: Slipway
The ramp for the slipway was designed and installed in the model. Once it had been established, the winch deck was constructed around the shore end of the slipway. Framing for the winch deck was not included in the original framing plans. As a result, in the Revit model, the winch deck was built in as a concrete structure. In future revisions of the model, the design of the winch deck would ideally be revised.

DESIGN SHEETS

The Revit outputs are collated in Book 2 of this report. The results have been categorised under the following headings:

S series
Structural plans
Structural elevations
Design details

A series
Main structure:
Architectural plans
Architectural elevations

STRUCTURAL DETAILS

CONNECTION OF CASTELLATED ROOF BEAM TO COLUMN

The curve of the arched castellated beam is an arc and less than 50% of the circumference of a circle. The intercept of the radius with the ends of the arch, therefore, creates an end-profile that does not correspond with the horizontal top end of the column to which it is to be connected. The contact end of the arched beam is therefore cut as shown in the diagram below, and a horizontal base-plate is attached, to allow face-to-face contact with the cap-plated head of the column.

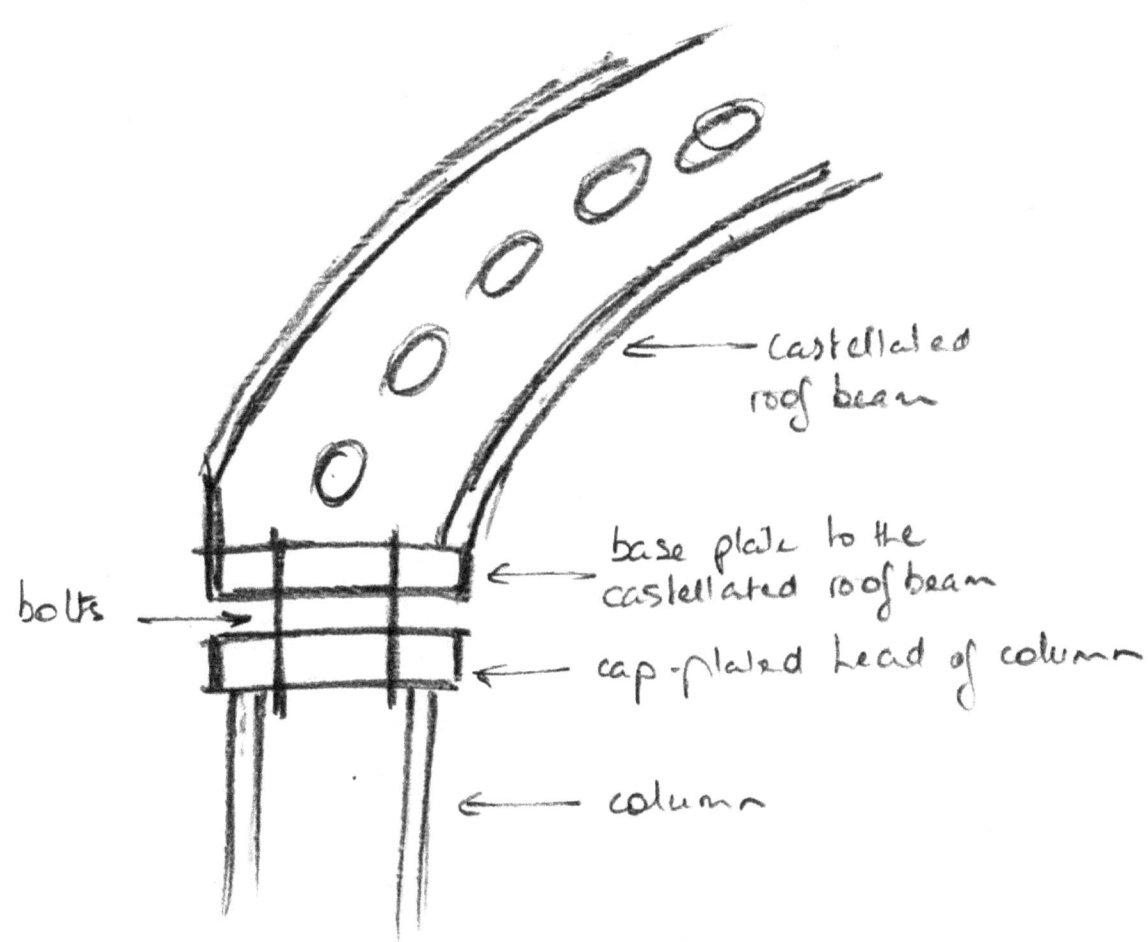

Figure 19: Connection of castellated roof beam to column

CONNECTION OF BRACE TO SUBSTRUCTURAL COLUMN

The braces are designed as circular hollow beams. The multiple connections at a single point required by cross-bracing are accommodated, as shown in the diagram below. A fin plate, with a surface area large enough to accommodate the connection of three brace elements, is bolted on to the substructure column. Three brace elements can then be run on to the fin plate.

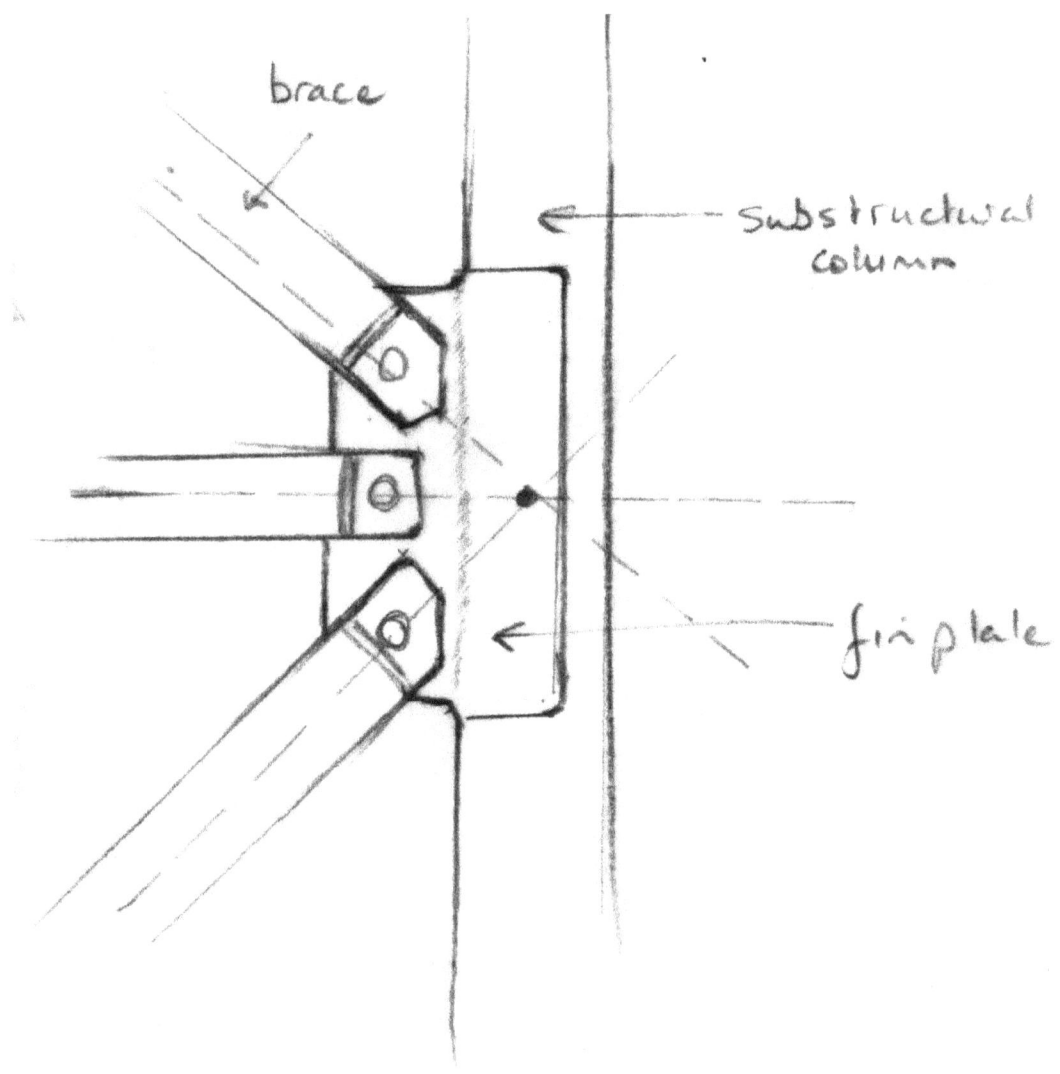

Figure 20: Connection of brace to substructure column

GANTRY CRANE

There is a requirement in the design brief for an overhead travelling crane that can lift at any position in the 21x 6m boat envelope. Our design advisors suggested that a gantry crane would serve the necessary purpose.

The Design Team has prepared draft drawings for the crane (below). The runway beam, the gantry beam, and the crane have been installed in the Revit model. In a future stage of the design process, the design of the gantry crane will be developed in greater detail.

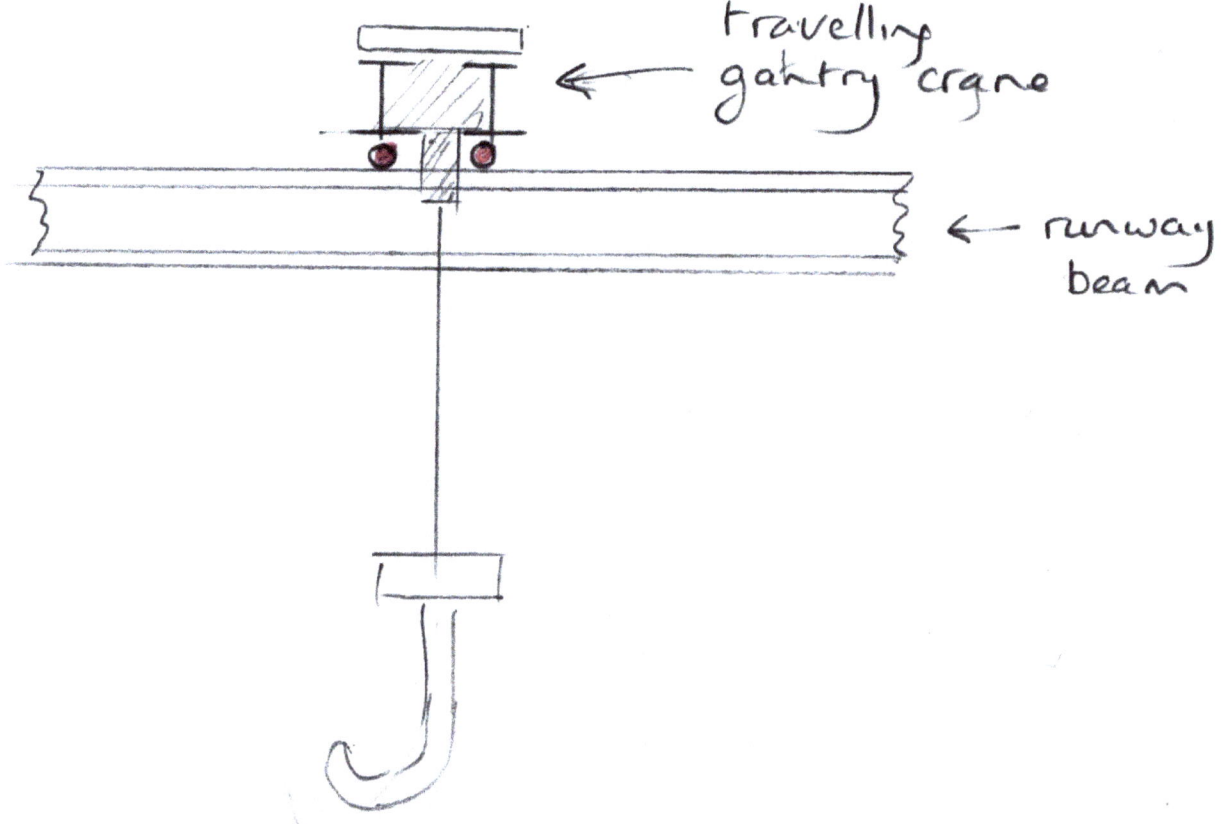

Figure 21: Runway beam

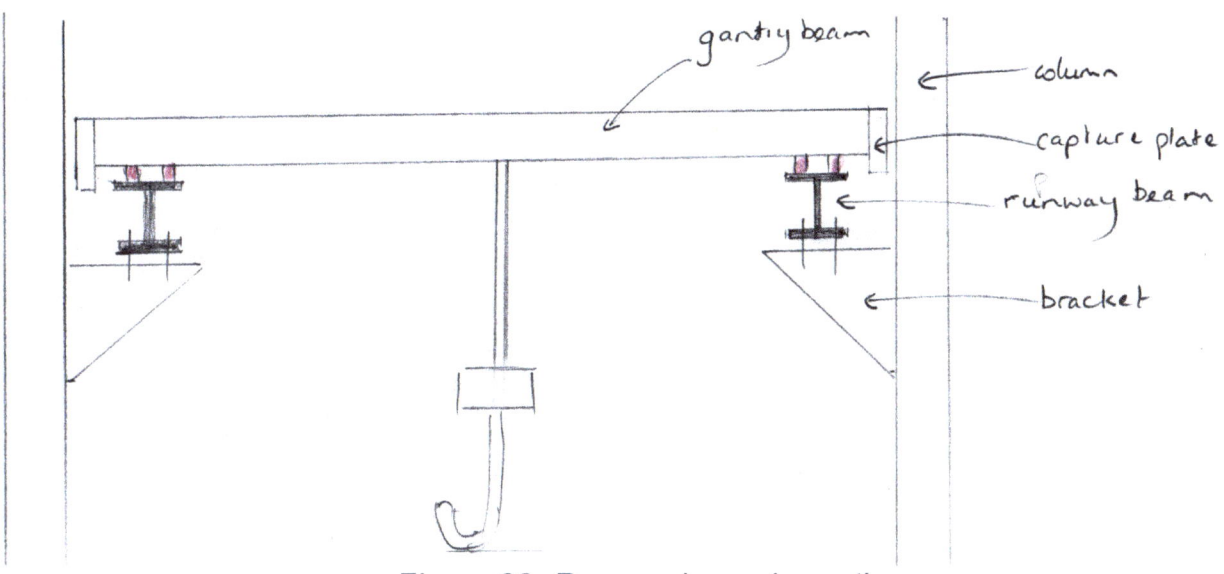

Figure 22: Runway beam in section

STRUCTURAL CALCULATIONS

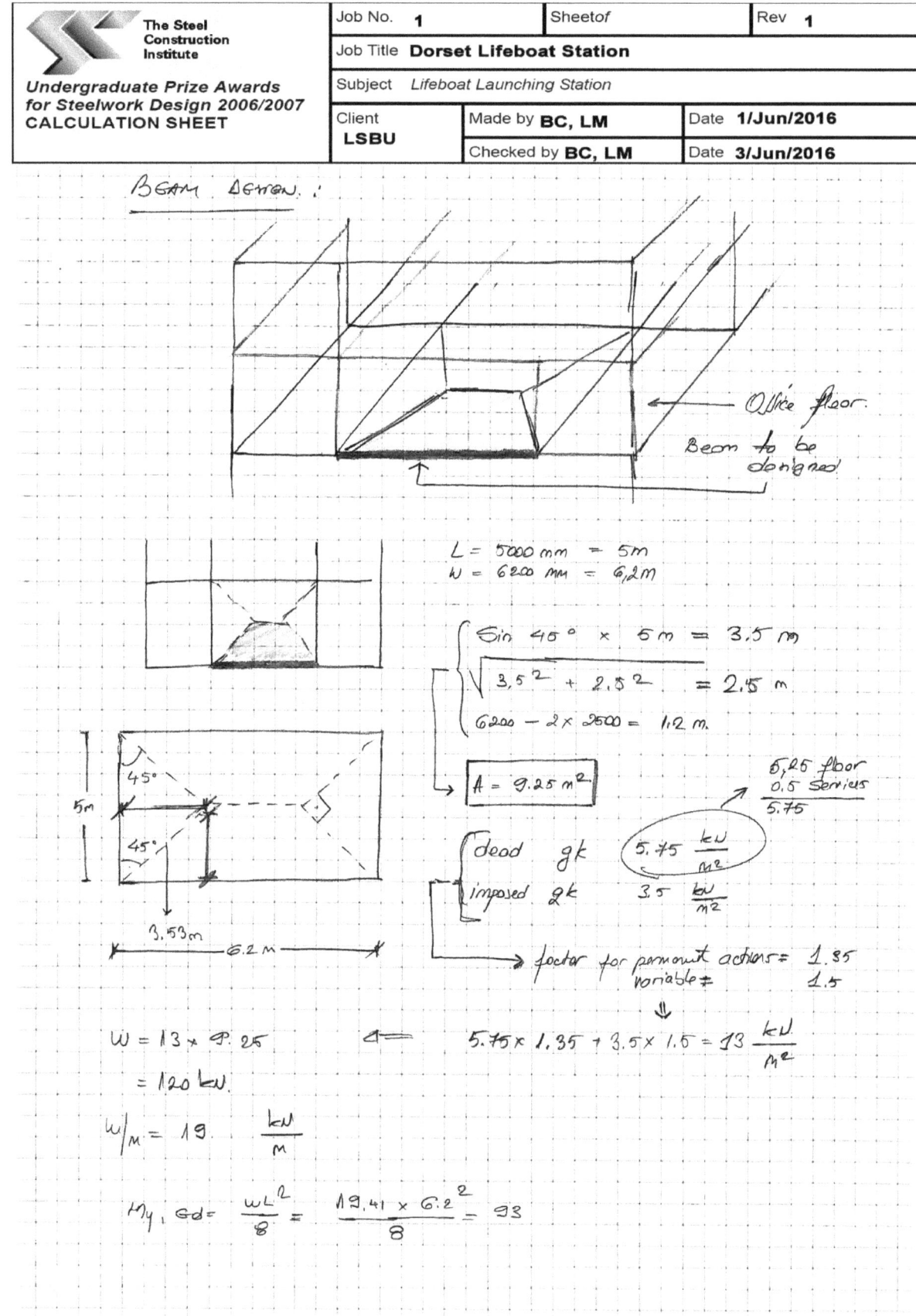

BEAM DESIGN:

$L = 5000\ mm = 5m$
$W = 6200\ mm = 6.2m$

$\sin 45° \times 5m = 3.5 m$
$\sqrt{3.5^2 + 2.5^2} = 2.5\ m$
$6200 - 2 \times 2500 = 1.2\ m$

$A = 9.25\ m^2$

$\dfrac{5.25\ \text{floor}}{0.5\ \text{services}}$
$\overline{5.75}$

dead g_k : 5.75 $\frac{kN}{m^2}$
imposed q_k : 3.5 $\frac{kN}{m^2}$

→ factor for permanent actions = 1.35
variable = 1.5

$5.75 \times 1.35 + 3.5 \times 1.5 = 13\ \frac{kN}{m^2}$

$W = 13 \times 9.25 = 120\ kN$

$W/m = 19\ \frac{kN}{m}$

$M_{y,Ed} = \dfrac{wL^2}{8} = \dfrac{19.41 \times 6.2^2}{8} = 93$

Shear buckling:

$$\frac{h_w}{t_w} \leq 72 \frac{\varepsilon}{\eta}$$

$$\frac{h_w}{t_w} = \frac{283}{7} = 39$$

$$72 \frac{\varepsilon}{\eta} = 72 \times \frac{0.9}{1.} = 66$$

No shear buckling check

Moment resistance.

$$\frac{M_{Ed}}{M_{c,Rd}} \leq 1$$

$$M_{c,Rd} = M_{pl,Rd} = \frac{W_{pl,y} \times f_y}{\gamma_{M0}} = \frac{539 \times 275}{1.} \times 10^{-3} = 148 \text{ kNm}$$

$$\frac{M_{y,Ed}}{M_{c,Rd}} \leq 1 \qquad \boxed{\frac{93}{148} = 0.62 < 1.} \text{ — adequate}$$

Deflection check:

$$\Delta = \frac{5 \times L^4 \times Q \times gk}{384 \times E \times I y} = \boxed{11 \text{ mm}}$$

L⁴ → 6200, Q → 1.5, gk → 5.75
E → 210×10³, Iy → 7171×10⁴

$$\frac{6200}{360} = \boxed{17}$$

11 < 17
OK
adequate ✓

 **The Steel Construction Institute** **Undergraduate Prize Awards for Steelwork Design 2006/2007 CALCULATION SHEET**	Job No. **1**	Sheet of		Rev **1**
	Job Title **Dorset Lifeboat Station**			
	Subject *Lifeboat Launching Station*			
	Client **LSBU**	Made by **BC, LM**		Date **1/Jun/2016**
		Checked by **BC, LM**		Date **3/Jun/2016**

Column design:

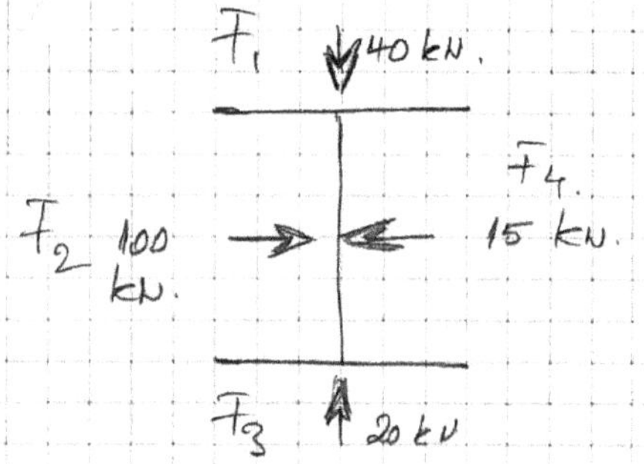

Pre-calculated reactions were obtained in a previous stage of the project.

The following 4 forces are acting on column which we are designing as per bellow.

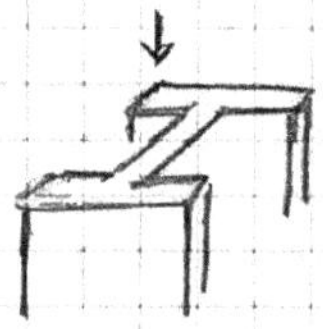

$F_1 = 40$ kN ↓
$F_2 = 100$ kN →
$F_4 = 15$ kN ←
$F_3 = 20$ kN ↑

Also, an axial loading of 800 kN has been calculated prior to this design exercise.

$$M_{z,Ed} = \left(\frac{100\,kN - 15\,kN}{2}\right) \times \left(\frac{0.2064}{2} + 0.1\right) = \boxed{8.636}$$

$$M_{y,Ed} = \frac{40\,kN - 20\,kN}{2} \times \left(\frac{0.2158}{2} + 0.1\right) = \boxed{2.079}$$

TRIAL SECTION CONSIDERED for eligibility:
UC 203×203×71

$h = 215.8\,mm = 0.2158\,m$
$r = 10.2\,mm$
$A = 904\,cm^2$
$W_{pl,y} = 799\,cm^3$
$W_{pl,z} = 374\,cm^3$
$i_y = 9.18\,cm$
$i_z = 5.3\,cm$
$t_f = 17.3\,mm$
$t_w = 10\,mm$
$b = 206.4\,mm = 0.2064\,m$

Yield strength (f_y)

Steel grade = S275
$t = 40\,mm$

$$\boxed{f_y = 275\,N/mm^2}$$

Section length:

Buckling length about $\begin{cases} y-y \text{ axis}: L_{cr,y} = 8.5\,m \\ z-z \text{ axis}: L_{cr,z} = 8.5\,m \end{cases}$

Section classification:

$$\varepsilon = \sqrt{\frac{235}{f_y}} = \sqrt{\frac{235}{275}} = \boxed{0.92}$$

① flange under compression (uniform)

$c = (b - t_w - 2r)/2 = (206.4 - 10 - 2 \times 10.2)/2 = 88\,mm$

$c/t_f = 88/17.3 = 5.09$

Limit value for class 1 ⇒ $\frac{c}{t_f} \leq 9\varepsilon$; $9 \times 0.92 = 8.28$ ⇒ $5.09 < 8.28$ ⇒ Compression ⇒ $\boxed{\text{class 1}}$

Job No. 1	Sheet of	Rev 1
Job Title **Dorset Lifeboat Station**		
Subject *Lifeboat Launching Station*		
Client **LSBU**	Made by **BC, LM**	Date **1/Jun/2016**
	Checked by **BC, LM**	Date **3/Jun/2016**

②. Internal compression – web under bending $c = d = 160.8$

limit for class 1: $\Rightarrow c/t_w = 160.8/10.0 = 16.8$

$c/t_w = 72\varepsilon = 33 \times 0.92 = 30.36$
$\Rightarrow 16.08 < 30.36$

⎫
⎬ class I
⎭

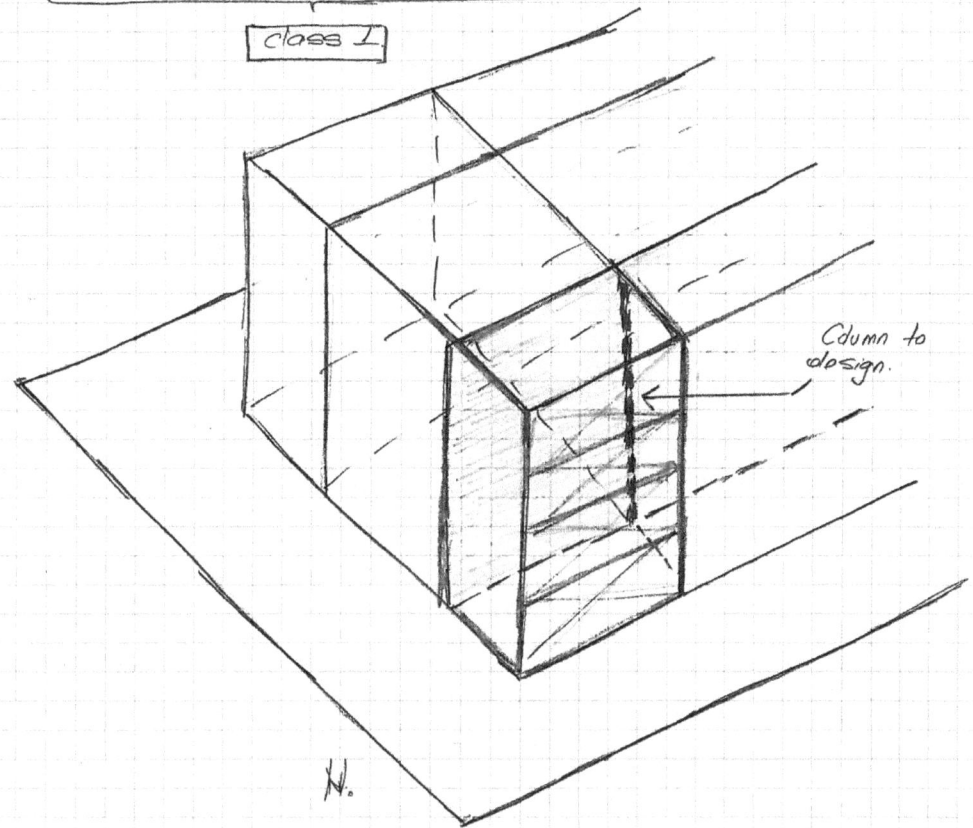

Column to design.

Buckling resistance:

$\lambda_1 = \pi \sqrt{\dfrac{E}{f_y}} = 93.9\,\varepsilon = 93.9 \times 0.92 = 86.39$

For y axis

$\overline{\lambda_y} = \dfrac{L_{cr}/i_y}{\lambda_1} = \dfrac{4290/9.18 \times 10}{86.39} = 0.541$

$\dfrac{h}{b} = \dfrac{215.8}{206.4} = 1.05 < 1.2$

&

$t_f = 17.3 < 100$

⎫
⎬ Buckling curve B for y axis
⎭

___graph___

$\eta_y = 0.81$

Job No.	1	Sheet of	Rev	1
Job Title	**Dorset Lifeboat Station**			
Subject	*Lifeboat Launching Station*			
Client **LSBU**	Made by **BC, LM**		Date	**1/Jun/2016**
	Checked by **BC, LM**		Date	**3/Jun/2016**

$$N_{by,ed} = \frac{\chi_y \times A \times f_y}{\gamma_{M_1}} = \frac{0.81 \times 90.4 \times 10^2 \times 275}{1} \times 10^{-3} = \boxed{2013.66 \text{ kN}}$$

For Z axis: $\bar{\lambda}_z = \frac{L_{cr}/i_z}{\lambda_1} = \frac{4290/5.3 \times 10}{86.39} = 0.937$;

$\frac{h}{b} = 1.05 < 1.2$

$t_f = 17.3 < 100 \rightarrow$ buckling curve C } $\chi_z = 0.66$
for Z axis

$$N_{bz,Rd} = \frac{\chi_z \times A \times f_y}{\gamma_1} = \frac{0.66 \times 90.4 \times 10^2 \times 275}{1} \times 10^{-3} = \boxed{1640.76 \text{ kN}}$$

$\boxed{N_{bz,Rd}} < \boxed{N_{by,Rd}}$ } $N_{bz,Rd} = 1640.4 \text{ kN}$
  1640   2013

Lateral buckling resistance:

y axis:

$\bar{\lambda}_1 = 0.9 \bar{\lambda}_y = 0.9 \times 0.541 = 0.49$

Rolled welded sections

$$\chi_{y_T} = \frac{1}{\phi_{y_T} + \sqrt{\phi_{y_T}^2 - \beta \bar{\lambda}_{y_T}^2}}$$

$\bar{\lambda}_0 = 0.4$

$\beta = 0.75$

$\alpha = 0.34$ — $\frac{h}{b} \leq 2. \rightarrow$ curve (b)

$\phi = 0.5[1 + \alpha_0(\bar{\lambda}_T - \bar{\lambda}_0) + \beta \bar{\lambda}_T^2] =$
$= 0.5[1 + 0.34(0.49 - 0.4) + 0.75 \cdot 0.49^2] = 0.61$

$\chi_T = \frac{1}{0.61 + \sqrt{0.61^2 - 0.75 \times 0.49^2}} = 0.95$

$\chi_T \leq 1 \longrightarrow 0.95 \leq 1 \text{ ok}$

$\chi_T \leq \boxed{\frac{1}{\bar{\lambda}_T^2}} \boxed{\frac{1}{0.49^2}} = 4.16 \rightarrow 0.95 < 4.16 \text{ ok}$

$$M_{by,Rd} = \frac{\chi_T \times W_{ply} \times f_y}{\gamma_M} = \frac{0.95 \times 799 \times 275}{1} \times 10^{-3} =$$

$$= \boxed{208.74 \text{ kNm}}$$

Job No. 1	Sheet of	Rev 1
Job Title **Dorset Lifeboat Station**		
Subject *Lifeboat Launching Station*		
Client **LSBU**	Made by **BC, LM**	Date **1/Jun/2016**
	Checked by **BC, LM**	Date **3/Jun/2016**

For z axis:

$\bar{\lambda}_T = 0.9 \, \bar{\lambda}_z = 0.9 \times 0.937 = 0.84$

$\chi_{zT} = \dfrac{1}{\phi_T + \sqrt{\phi_T^2 - \beta \bar{\lambda}_T^2}}$ $\quad \begin{cases} \bar{\lambda}_0 = 0.4 \\ \beta = 0.75 \\ \alpha_0 = 0.34 \end{cases}$ $\quad \alpha_0 \to \dfrac{h}{b} \leq 2 \Rightarrow$ curve b.

$\phi_T = 0.5 \left[1 + \alpha_0 (\bar{\lambda}_T - \bar{\lambda}_0) + \beta \bar{\lambda}_T^2 \right] =$

$0.5 \left[1 + 0.34 (0.84 - 0.4) + 0.75 \times 0.84^2 \right] = 0.84$

$\chi_{zT} = \dfrac{1}{0.84 + \sqrt{0.84^2 - 0.75 \times 0.84^2}} = 0.79$

$\chi_{zT} \leq 1 \longrightarrow 0.79 < 1 \; \text{ok}.$

$\chi_{zT} \leq \dfrac{1}{\bar{\lambda}_{zT}^2} \longrightarrow 0.79 < 1.4 \; \text{ok}$

$Mb_{zRd} = \dfrac{\chi_{zT} \times W_{pl} \times f_y}{1}$

$= \dfrac{0.79 \times 374 \times 275}{1} \times 10^{-3} = \boxed{81.25 \; kNm}$

Analysis on bending & axial buckling:

$$\underbrace{\dfrac{N_{ed}}{N_{bz,Rd}}}_{\substack{800 \, kN \\ 1640 \, kN}} + \underbrace{\dfrac{M_{y,ed}}{M_{by,Rd}}}_{\substack{2.049 \\ 208.74 \, kNm}} + 1.5 \underbrace{\dfrac{M_{z,ed}}{M_{bz,Rd}}}_{\substack{8.636 \\ 81.25 \, kNm}} \leq 1$$

$$\downarrow$$
$$0.65 \quad\quad\quad\quad \leq 1$$

SECTION IS ADEQUATE

Wind and wave force – analysis:

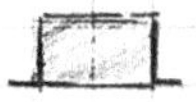

wind → $1.5 \, kN/m^2$ × 1.5 (factor) × 12 m × 9.4 m = 253.8 kN
(wind force for main part of the building)

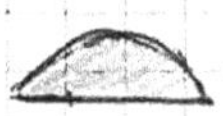

wind → $1.5 \, kN/m^2$ × 1.5 (factor) × 13.23 m^2 = 29.78 kN
(wind force for arched castillated beam) — area of arched castillated beam

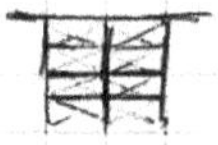

waves → $5 \, kN/m^2$ × 1.5 (factor) × 1.25 × 52.24 m^2 = 489.77 kN

dinamic nature of forces (further factor × 1.25)

Force per building impact by air + waves:

Area exposed to waves:

	#	length	Σ	thickness	Σ area
braces	16	6.44	103.01	0.2 m	20.60 m^2
vertical elements	4	17.6	37.6	0.35 m	13.16 m^2
horizontal elements	3	9.4	52.8	0.35 m	18.48 m^2

52.24 m^2

Wind force acting per bracing:

$$253.8 \, kN + 29.78 \, kN = 311.94 \, kN$$

$$\frac{311.94}{4 \text{ (systems of bracing)}} = 77.9841 \, kN / 12 \text{ (m height)} = 6.5 \, kN/m \text{ wind load}$$

Wave force acting per bracing:

$$\frac{489.77 \, kN}{4} = 122.44 \, kN / 17.6 \text{ (height under water)} = 6.96 \, kN/m \text{ wave load}$$

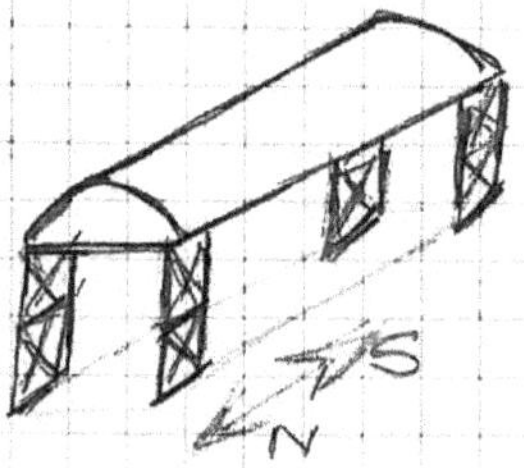

For simplicity, we considered that the water level will reach main platform.

$6.5 \, \frac{kN}{m}$ N-S wind load

$6.96 \, \frac{kN}{m}$ N-S wave load

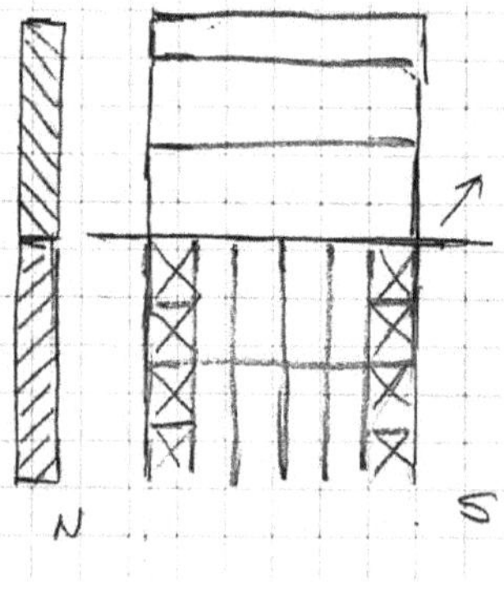

main platform

	The Steel Construction Institute	Job No. 1	Sheet of	Rev 1
Undergraduate Prize Awards for Steelwork Design 2006/2007 CALCULATION SHEET		Job Title **Dorset Lifeboat Station**		
		Subject *Lifeboat Launching Station*		
		Client **LSBU**	Made by **BC, LM**	Date **1/Jun/2016**
			Checked by **BC, LM**	Date **3/Jun/2016**

FORCES GENERATED BY WIND, WAVES:

Bracing systems designed resist against wind & wave forces acting N–S. There is a total of 4 bracing systems.
The wind & wave impact has been assessed per single bracing system:

				height above ground level
F_{roof}	1.75 m × 6.5 kN	= 11.37 kN	at 29.6 m	
F_6	4 m × 6.5 kN	= 25.99 kN	26.1 m	
F_5	4.25 m × 6.5 kN	= 27.62 kN	21.6 m	
F_4	2 m × 6.5 kN + 2.2 × 6.96 (m)(kN)	= 28.30 kN	17.6 m	
F_3	4.4 × 6.96 (m)(kN)	= 30.61 kN	13.2 m	
F_2	4.4 × 6.96 (m)(kN)	= 30.61 kN	8.8 m	
F_1	4.4 × 6.96 (m)(kN)	= 30.61 kN	4.4 m	
F_{ground}	2.2 × 6.96 (m)(kN)	= 15.31 kN	0	
		200.43 kN		

● $H_a + H_t = \boxed{200.43}$ kN

11.37 + 26 + 27 + 38.3 + (30.6) × 3 + 15.3

● $V_t = (11.34 × 29.6 +$
   $26 × 26 +$
   $27.6 × 21.6 +$
   $28.6 × 17 +$
   $30.6 × 13.2 +$
   $30.6 × 8.8 +$
   $30.6 × 4.4) / \boxed{5m} = 583$ kN
   (length)

● $V_A = 583$ kN $(V_A + V_t = 0)$

● $H_A = 15.31$ kN $(H_A + F_g = 0)$

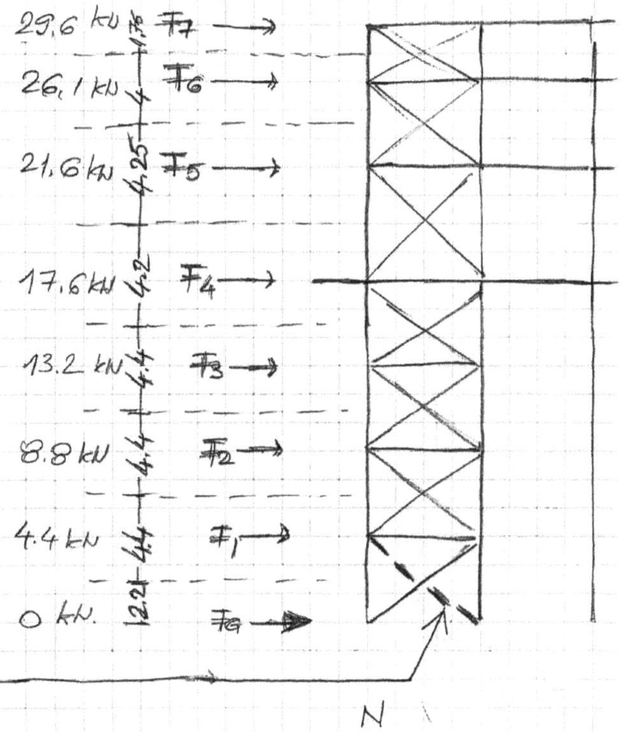

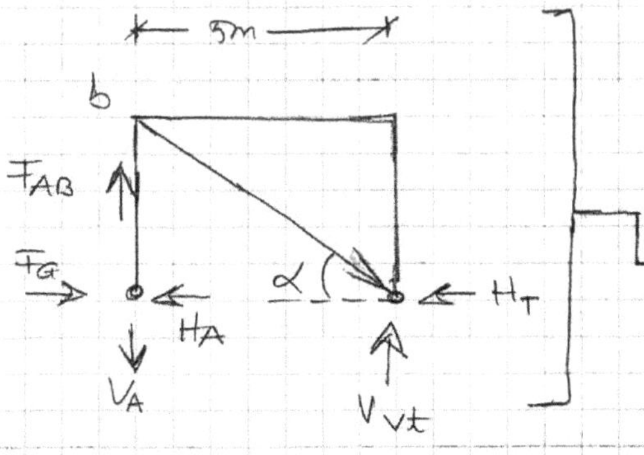

- $H_t = 200.43 + \boxed{15.31}^{HA} = 215.73\ kN$ $AB = 4.4$
- $F_{AB} = 583\ kN. \longrightarrow F_A + F_{AB} = 0.$ $AT = 5$
- $F_{BT} = \boxed{215}^{HT}/\cos(41.35) = 287.37\ kN$ Hypoteneuse $= 6.66$

  $\sin \angle = 0.66$

  Angle $= 41.35°$

$F_{BT} = 287.87$ $\}$
$\ell = 6.6\ m.$

BRACING DESIGN BASED ON THE ABOVE

$140 \times 140 \times 5 \longrightarrow 294\ kN$
$L = 7m$

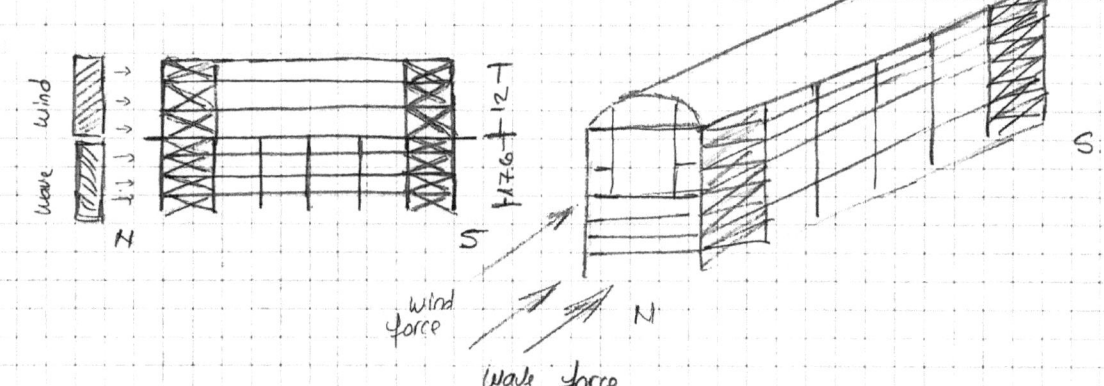

The brace designed above is located on first support structure as per bellow, towards N side.

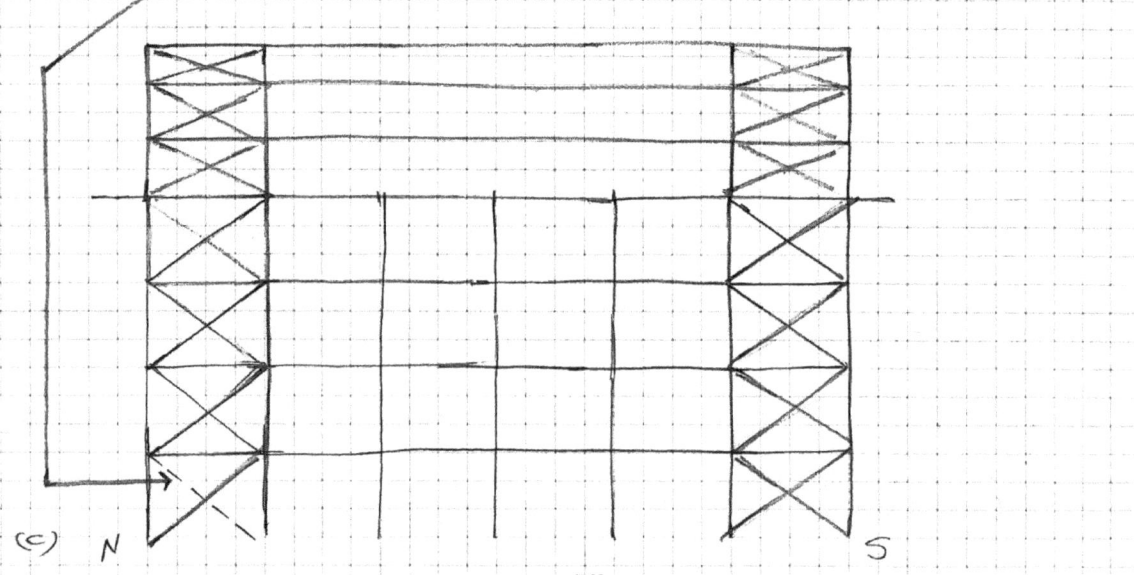

		Job No. 1	Sheet of	Rev 1
 The Steel Construction Institute		Job Title **Dorset Lifeboat Station**		
Undergraduate Prize Awards for Steelwork Design 2006/2007 CALCULATION SHEET		Subject *Lifeboat Launching Station*		
		Client **LSBU**	Made by **BC, LM**	Date **1/Jun/2016**
			Checked by **BC, LM**	Date **3/Jun/2016**

Column – beam:

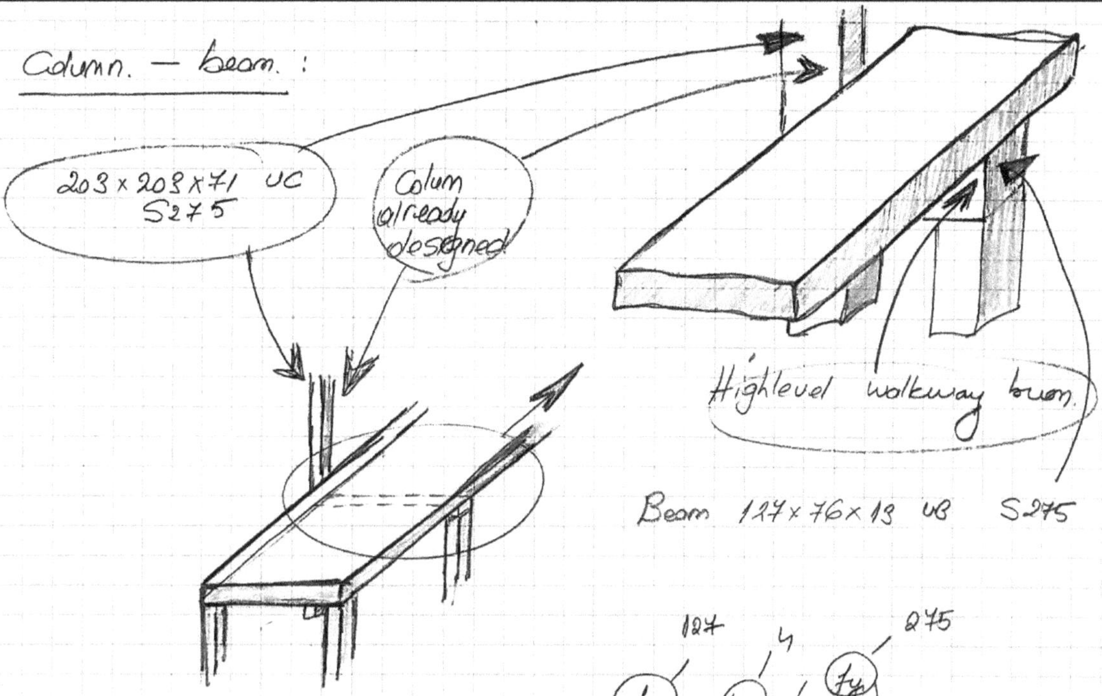

$203 \times 203 \times 71$ UC S275

Column already designed

High level walkway beam

Beam $127 \times 76 \times 13$ UB S275

Beam:

$f_y = 275 \; N/mm^2$

$f_u = 430 \; N/mm^2$

$h_w = 127 \; mm$

$t_w = 4 \; mm$

$t_f = 7.6 \; mm$

$$V_{o,Rd} \leq \frac{h_w \times t_w \left(\frac{f_y}{\sqrt{3}}\right)}{\gamma_{M0}} \times 10^{-3} = 80 \; kN$$

(127, 4, 275)

$8.44 \; kN < 0.75 \times 80 \; kN$

$\underbrace{\qquad\qquad}_{60 \; kN}$

$hb < 500 mm$

8 or 10 mm endplate proposed

End plate min $0.6 \; h_w = 0.6 \times 127 = 76.2$

80 mm end plate

Assuming M20 bolts

# of bolts $= \dfrac{V_{ed}}{74} = \dfrac{8.44}{74} = 0.12$ } 4× M20

Self weight has been ignored

Bolt details: treated bolts: M20 grade 8.8 50mm

$A_s = 84 \; mm^2$
$d_o = 14 \; mm$
$d_w = 24$
$f_{ub} = 640$
$f_{ub} = 800$

Locations & Spacing:

	min	$d_0 = 2.2 \times 14 = 30.8$	ok
$p_1 = 40\,mm$	2.2		
$e_1 = 25\,mm$	1.2	$1.2 \times 14 = 16.8$	ok
$e_2 = 30\,mm$	1.2	$1.2 \times 14 = 16.8$	ok
$p_2 = 40\,mm$	2.4	$2.4 \times 14 = 33.6$	ok

Weld design

$$a \geq 0.39 \times t_w = 0.39 \times 4 = 1.56\,mm \Rightarrow 2\,mm$$

$leg = 6\,mm$

Bolts in shear:

$$F_{V,Rd} = \frac{d_v \times f_{ub} \times A}{\gamma_{M_2}} = \frac{0.6 \times 800 \times 84.3}{1.25} \times 10^{-3} = 32\,kN$$

4 bolts $V_{Rd} = 4 \times 32 = \boxed{130\,kN}$

END PLATE

$$F_{b,Rd} = \frac{k_1 \times \alpha_b \times f_{up} \times d \times t_p}{\gamma_{M_2}} \longrightarrow min\left(\alpha_d\,;\,\frac{f_{ub}}{f_{up}}\,;\,1.0\right)$$

$\alpha_d = \frac{e_1}{3d_0} \rightarrow$ end bolts

$\alpha_d = \frac{p_1}{3d_0} - \frac{1}{4} \rightarrow$ inner bolts

FOR END BOLTS:

$$\alpha_b = min\left(\frac{25}{3 \times 14}\,;\,\frac{800}{430}\,;\,1\right) = min\left(0.6\,;\,1.86\,;\,1\right) = 0.6$$

FOR INNER BOLT:

$$\alpha_b = min\left(\frac{40}{3 \times 14} - \frac{1}{4}\,;\,\frac{800}{430}\,;\,1\right) = min\left(0.7\,;\,1.86\,;\,1\right) = 0.7$$

$$k_1 = min\left(2.8\,\frac{e_2}{d_0} - 1.7\,,\,2.5\right) = min\left(2.8 \times \frac{30}{14} - 1.7\,;\,2.5\right) =$$
$$= min\left(4.3\,,\,2.5\right) = 2.5$$

END OF PLATE BOLTS:

$$F_{b,Rd} = \frac{2.5 \times 0.6 \times 430 \times 12 \times 10}{1.25} \times 10^{-3} = 61.92\,kN$$

$$F_{V,Rd} = \frac{2.5 \times 0.7 \times 430 \times 12 \times 10}{1.25} \times 10^{-3} = 72.24\,kN$$ adequate — ok

4 bolts $\longrightarrow V_{Pd,R} = 4 \times 61 = 247\,kN$

$\boxed{\begin{array}{cc} kN & kN \\ 69.86 & > 9 \end{array}}$

BEAM WEB IN SHEAR:

110×4

$$V_{pl,Rd} = V_{Rd,g} = \left[\frac{A_v(f_{yb})}{\sqrt{3}}\right] / \gamma_{M_0} \neq 0.9 = \boxed{69\,kN}$$

	Job No. 1	Sheet of	Rev 1
The Steel Construction Institute	Job Title **Dorset Lifeboat Station**		
Undergraduate Prize Awards for Steelwork Design 2006/2007 **CALCULATION SHEET**	Subject *Lifeboat Launching Station*		
	Client **LSBU**	Made by **BC, LM**	Date **1/Jun/2016**
		Checked by **BC, LM**	Date **3/Jun/2016**

BOLTS IN TENSION:

$$F_{t,Rd} = \frac{k_2 \times f_{ub} \times A_s}{\gamma_{M0}} \times 10^{-3} = 55.18 \text{ kN}$$

with $k_2 = 0.9$, $f_{ub} = 800$, $A_s = 843$, $\gamma_{M0} = 1.1$

For 4 bolts → $N_{Rd} = 4 \times 55.18 = 220$

END PLATE IN BENDING:

$$N_{Rd} = \min(F_{Rd,up1}, F_{Rd,up2})$$

$$\frac{(8m_p - 2e_w) M_{pl,1,Rd}}{2m_p n_p - e_w(m_p + n_p)} \qquad \frac{2M_{p,2,Rd} + n_p \sum F_{b,Rd}}{m_p + n_p}$$

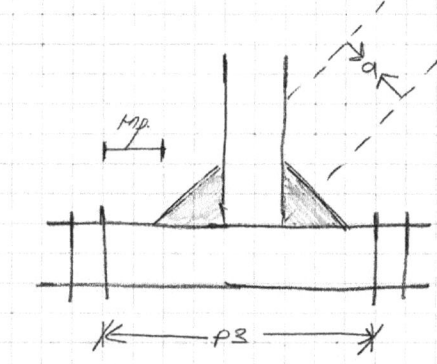

$n_p = \min(e_2, e_{2,c}, 1.25 m_p) =$
$= \min(30, 77.3, 19.67) = 19.67$

$m_p = \dfrac{p_3 - t_{w,b} - 2 \times 0.8 a \sqrt{2}}{2}$

$= \dfrac{40 - 4 - 2 \times 0.8 \times 2 \sqrt{2}}{2} = \boxed{15 \text{ mm}}$

$e_w = \dfrac{d_w}{4} = \dfrac{24}{4} = \boxed{6 \text{ mm}}$

$\sum L_{eff,1} = 90 \text{ mm}$
$\sum L_{eff,2} = 90 \text{ mm}$

$M_{pl,Rd,1} = M_{p,Rd,2} = \dfrac{1}{4} \dfrac{\sum L_{eff} \times t_p^2 \times f_{y,p}}{\gamma_{M0}} =$

$= \dfrac{1}{4} \times \dfrac{90 \times 10^2 \times 430}{1.1} \times 10^{-6} = 0.88 \text{ kNm}$

Node 1
$F_{T,Rd,1} = \dfrac{(8 \times 19 - 2 \times 6) \times 0.88 \times 10^3}{2 \times 15 \times 19 - 6(15 + 19)}$
$= 314 \text{ kN}$

Node 2
$F_{T,Rd,2} = \dfrac{(2 \times 0.88 \times 10^3) + (19 \times 270)}{15 \times 19}$
$= 22 \text{ kN}$

$V_{Rd} = 69 \text{ kN}$
$N_{Rd} = 22 \text{ kN}$

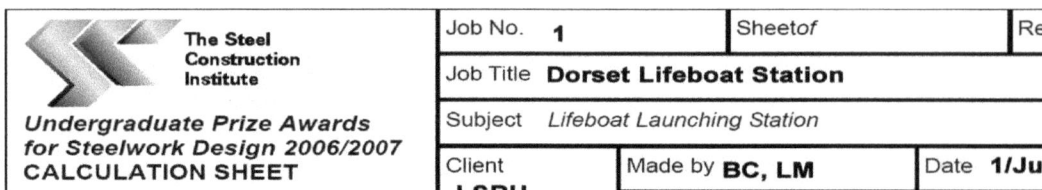

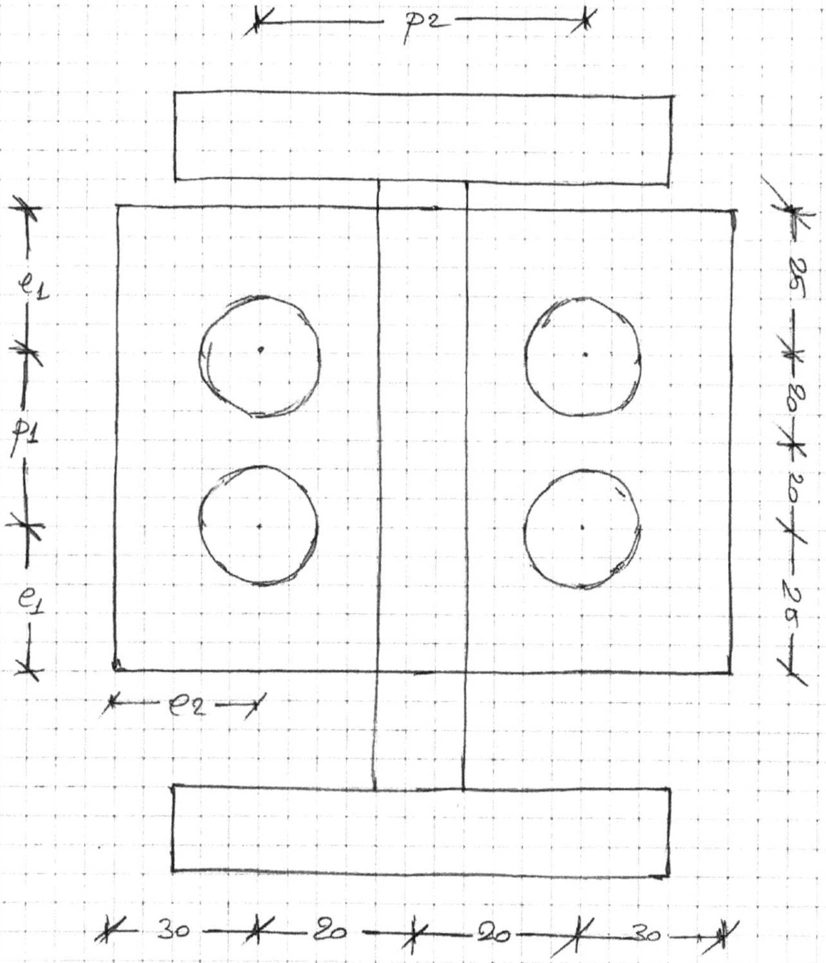

APPENDIX

DESIGN DEVELOPMENT

DESIGN SPECIFICATION	DESIGN PROCESS	DESIGN PRINCIPLE
Floor area of boat envelope: 6 x 20m	The floor area of the boat envelope is defined and laid out.	Design the structure around its primary function.
An internal high-level walkway to be provided down each side of boat envelope.	The floor area of the walkway is defined and laid out.	
	Horizontal grid lines are placed at the boundaries of each functional area. A central horizontal grid line is placed to accommodate the central keel of the boat. Vertical grid lines are placed at 5m intervals, thus establishing a series of 5m bays.	Establishing successive, equally-spaced bays of 5m along the x axis is consistent, firstly, with the principles of symmetry, and, secondly with the principle that repetitive processes are more cost effective.

Figure 23: Design development - the process (notes)

THE GRID

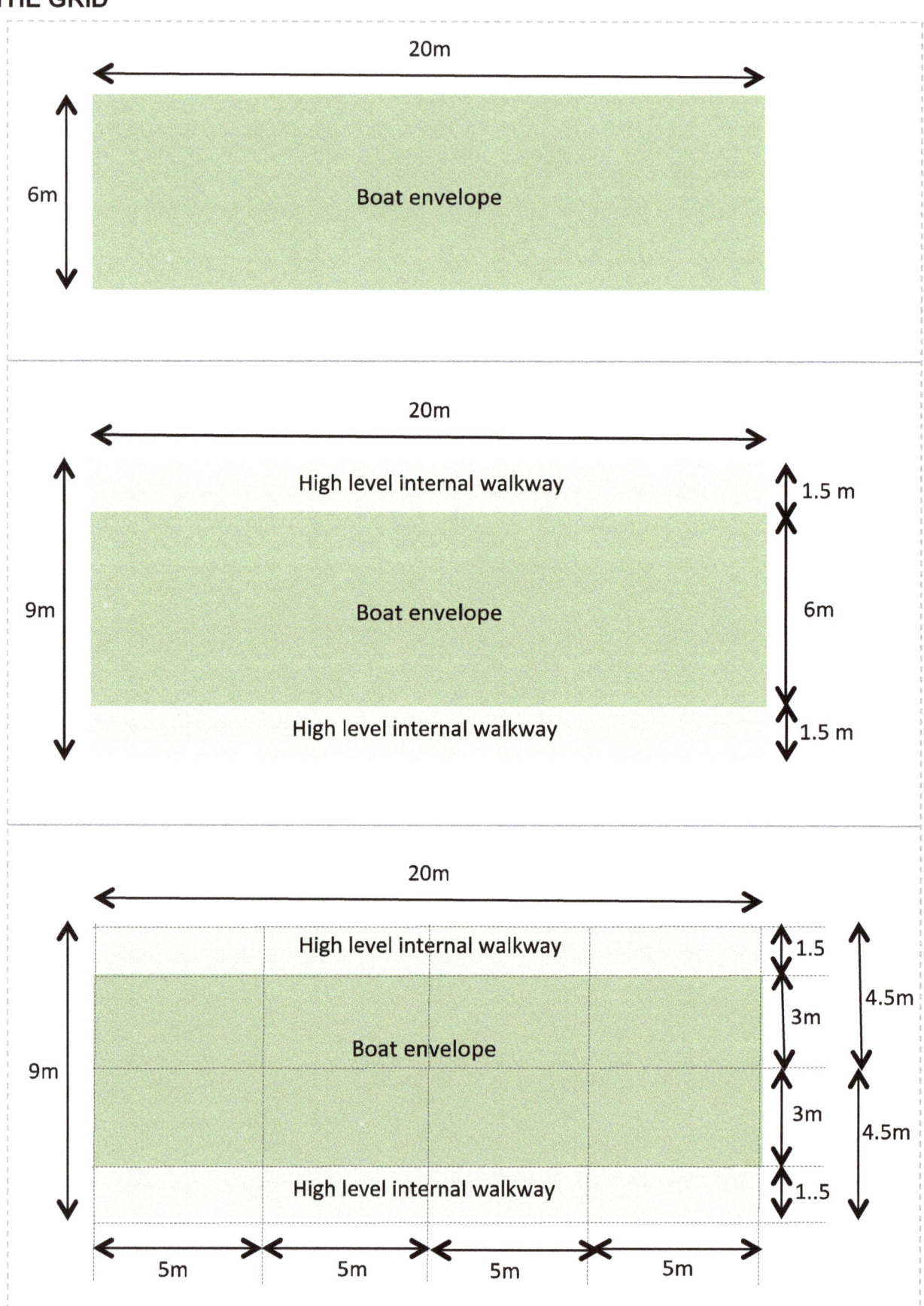

Figure 24: The grid - calculations

A winch is to be provided at the shore-end of the structure.	A 5 x 9m bay is added to the shore end of the grid, to provide a **service area** for personnel.	Maintaining the width of the new bay at 9m maintains the symmetry of the grid.
	The service area accommodates the winch and provides access to the boat for service personnel. Access to the high level walkway will arise from this bay.	
Access to the structure is to be provided by a foot-bridge at the shore end.	A 5 x 9m bay is added to the shore end of the grid, to serve as an **access area**.	Accommodating the access area as an additional bay on the horizontal axis is consistent with the principles of symmetry and of repetitivity.
	The access area provides access to the service area for personnel and to public areas for the public.	
An area of 100m2 is to be provided for office and mess.	The office and mess space is to be provided at an upper leve.	Space that is fundamental to the operational and access functions of the structure have been provided for.
	The 9 x 30m grid therefore establishes the footprint of the structure.	In line with the principle of minimising the footprint of the structure, space for other functions (office, mess, etc) will be provided for at an upper level.
	With the footprint now determined, the placing of supporting columns can be determined.	The major part of the vertical load imposed by the boat will be along the keel. This necessitates a row of columns along the central horizontal gridline.
	The location of supporting columns is shown by the black squares.	Placing a row of columns along the lower and upper gridlines would require a connecting beam span of 4.5m.

Figure 25: Design development - the process (continued notes)

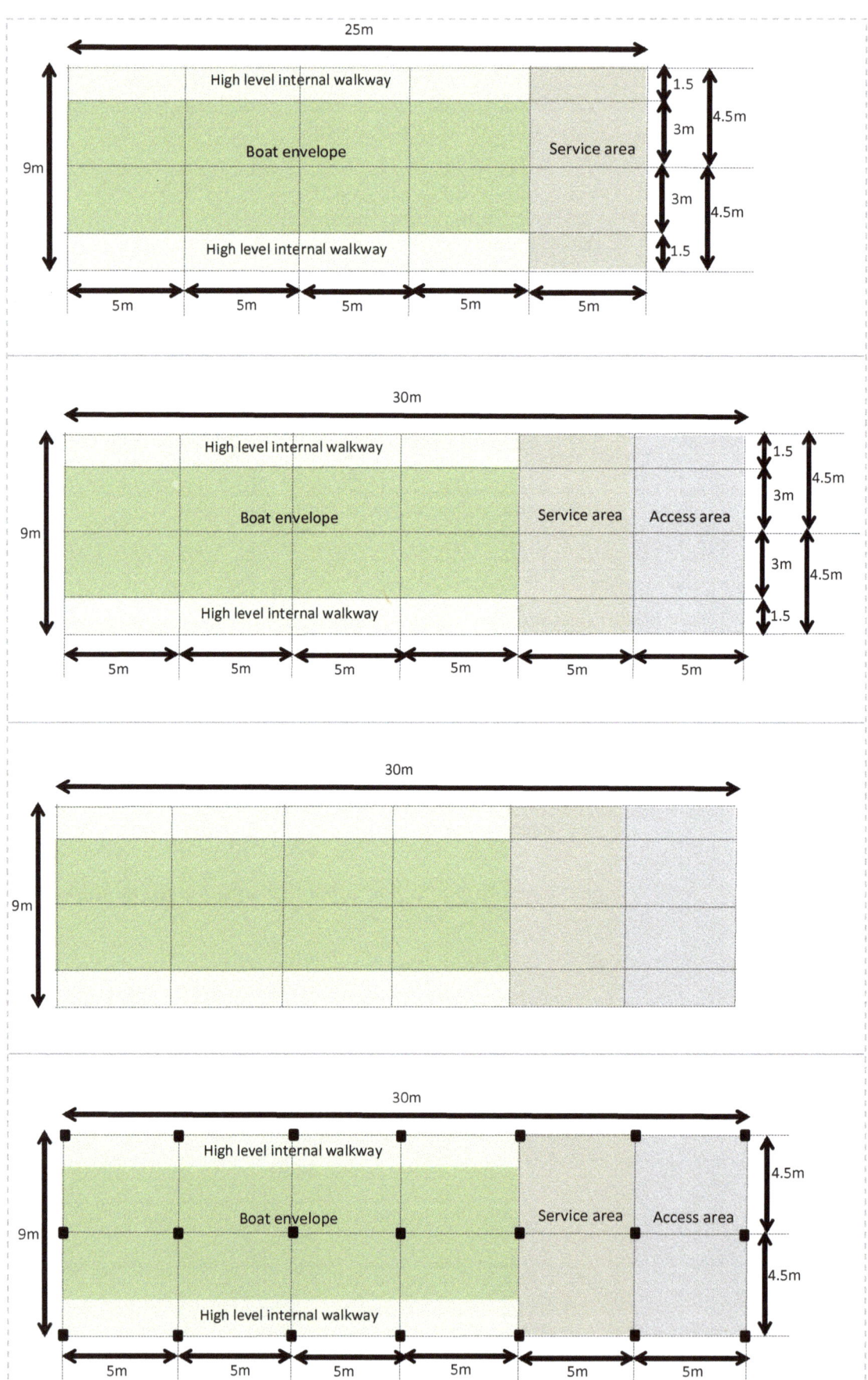

Figure 26: Calculations

An external walkway of 1.5m width is to be provided around the perimeter of the main structure.	Provision is made for a walkway.	Since the walkway is external and will not support any part of the main structure, it lies outside the footprint of the main structure. This arrangement suggests that support for the external walkway could be provided either by a cantilever arrangement or by a series of brackets.
	Removing the annotations shown in earlier diagrams reveals the grid and the location of the substructural columns.	In keeping with the principle of symmetry, the grid is symmetrical in both directions. In keeping with the principle of the standardisation of components, this grid represent only 4 beam lengths: 5.0m 4.5m 3.0m 1.5 m

Figure 27: Design development - the process (continued notes)

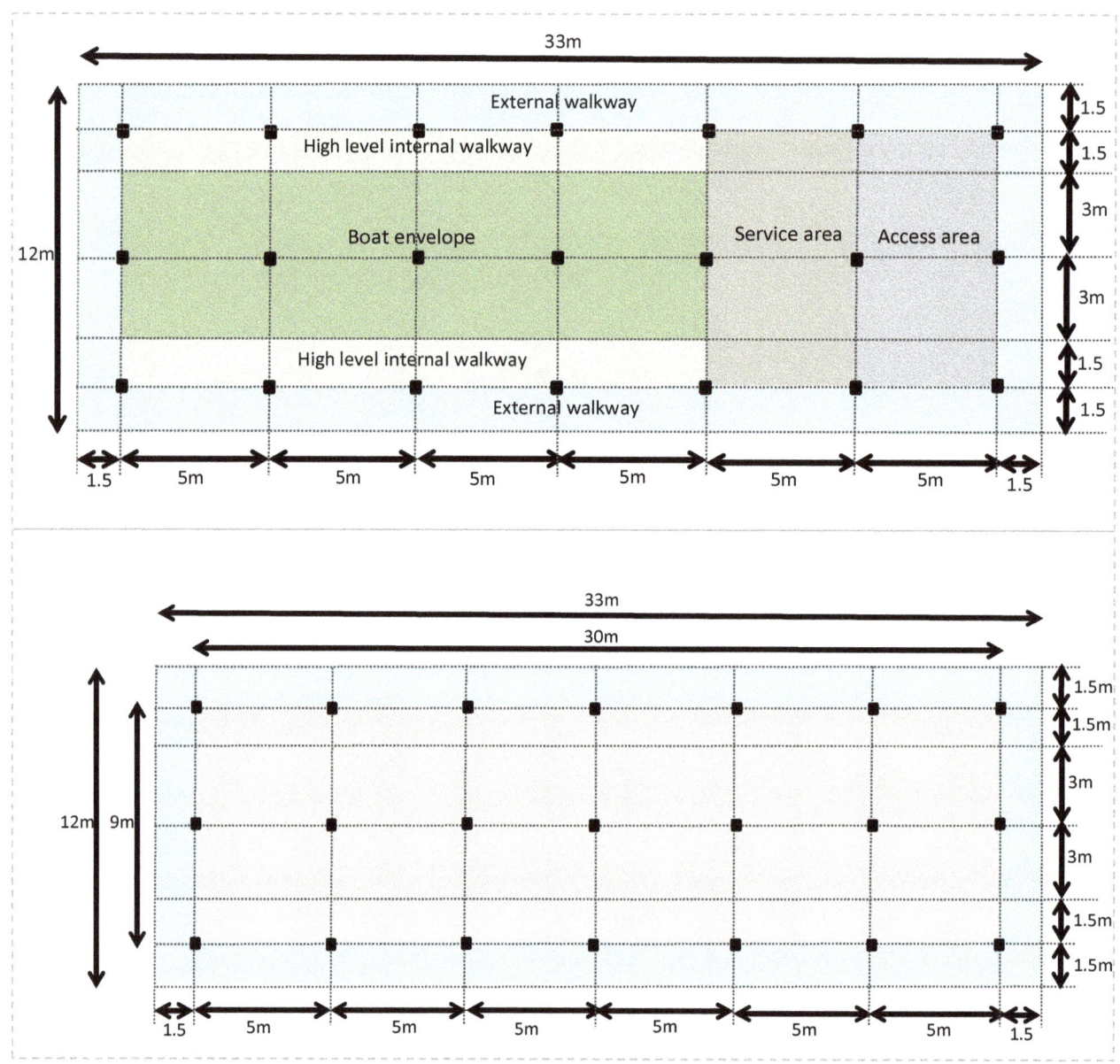

Figure 28: Calculations

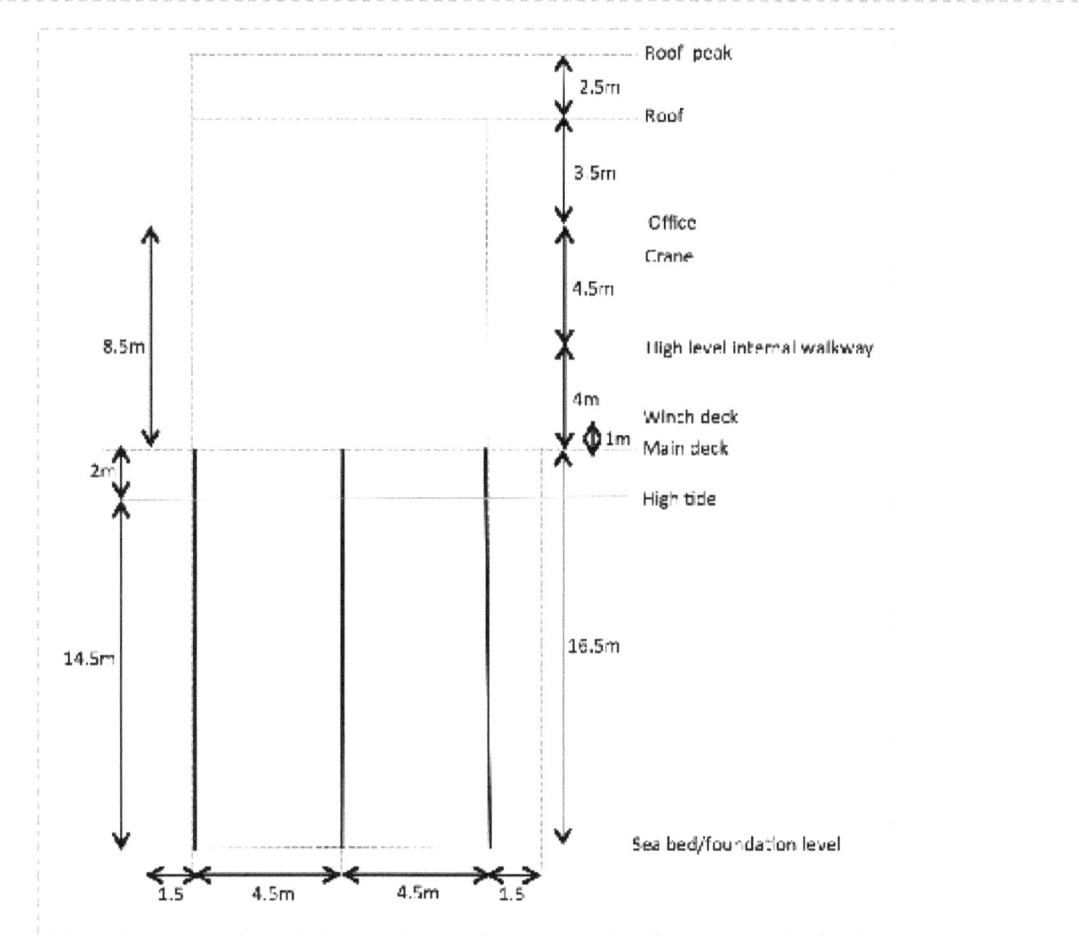

NORTH AND SOUTH ELEVATIONS

The process whereby the levels for the Revit model were generated is laid out

LEVEL	DESIGN SPECIFICATIONS
SUBSTRUCTURE	
MAIN DECK	*Minimum distance between maximum high water level and lower surface of main structure: 2m*
SLIPWAY	*A slipway is to be provided, on the basis of the dimensions shown in the diagram overleaf.*

Figure 29: Calculations north and south elevations

THE LEVELS

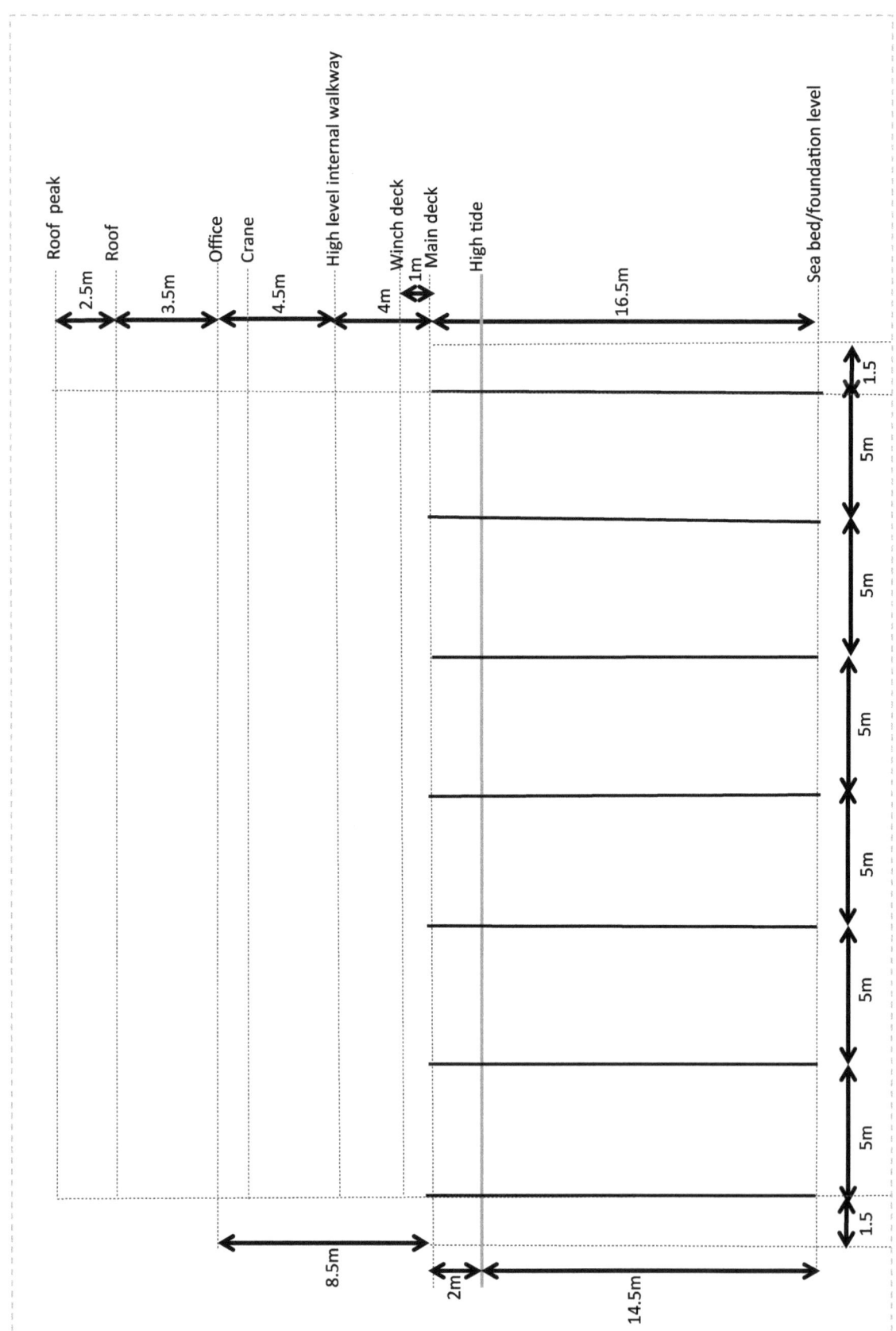

Figure 30: West and east elevations

WINCH PLATFORM	*At the shore end of the lifeboat station, a winch is located to haul the lifeboat back into the lifeboat house (as per the diagram below)*
HIGH-LEVEL INTERNAL WALKWAY	*A high-level internal walkway is to be provided at each side of life-boat at a height of 4m.*
OFFICE	*Office and mess space of 100m².*
ROOF	*Ceiling height of 3m at office level.*
ROOF PEAK	
CRANE	*Provision is to be made for an overhead travelling crane with a lift capacity of 2 tonnes that can lift anywhere in the 20l x 6m boat envelope.*

Figure 31: Winch

In terms of the specifications given the height of the winch deck above the main deck is calculated to be 0.67m. For the sake of convenience, this is rounded up to 1m.	
The office floor is placed 8.5m above the main deck. Height of the boat envelope is specified as 6m and a further 2.5m is provided, to allow for installation of a gantry crane.	In line with the principle of minimising the footprint of the structure, office space is located above the operational level.
Roof level is placed 3.5m above the office level, with 0.5m being provided for ceiling and services within.	
Roof peak level is placed 2.5m above roof level.	Comment: ultimately 2.5m proved to be over-generous. A decision was made during the design development process to eliminate a ceiling to the office level, with the vaulted roof itself serving as the ceiling. As a result, the office height is 6m at the centre. The design would therefore have to be modified at future stages of its development.
The crane level is placed 1.5m below the office level.	Comment: the distance of 1.5m is inadequate, since it is not sufficient allow access for service personnel. In future stages of the design development, the distance would have to be increased.

Figure 32: Technical details

LEVEL	DESIGN PROCESS
MAIN DECK	Location of substructural columns has been established as per the previous section. The structural envelope was established, differentiating between internal space and the external walkway.
	Programming has been established, as per Section xxx. A 6m stretch of external walkway is removed at south end to allow access for boat.
	Beams are placed in relation to established substructural column positions. Edge beams to support external walkway are indicated in blue. Supporting beams for the walksway are not provided, since a decision has been made to support the walkway on gallows brackets.

Figure 33: Design of the main deck

FRAMING PLANS

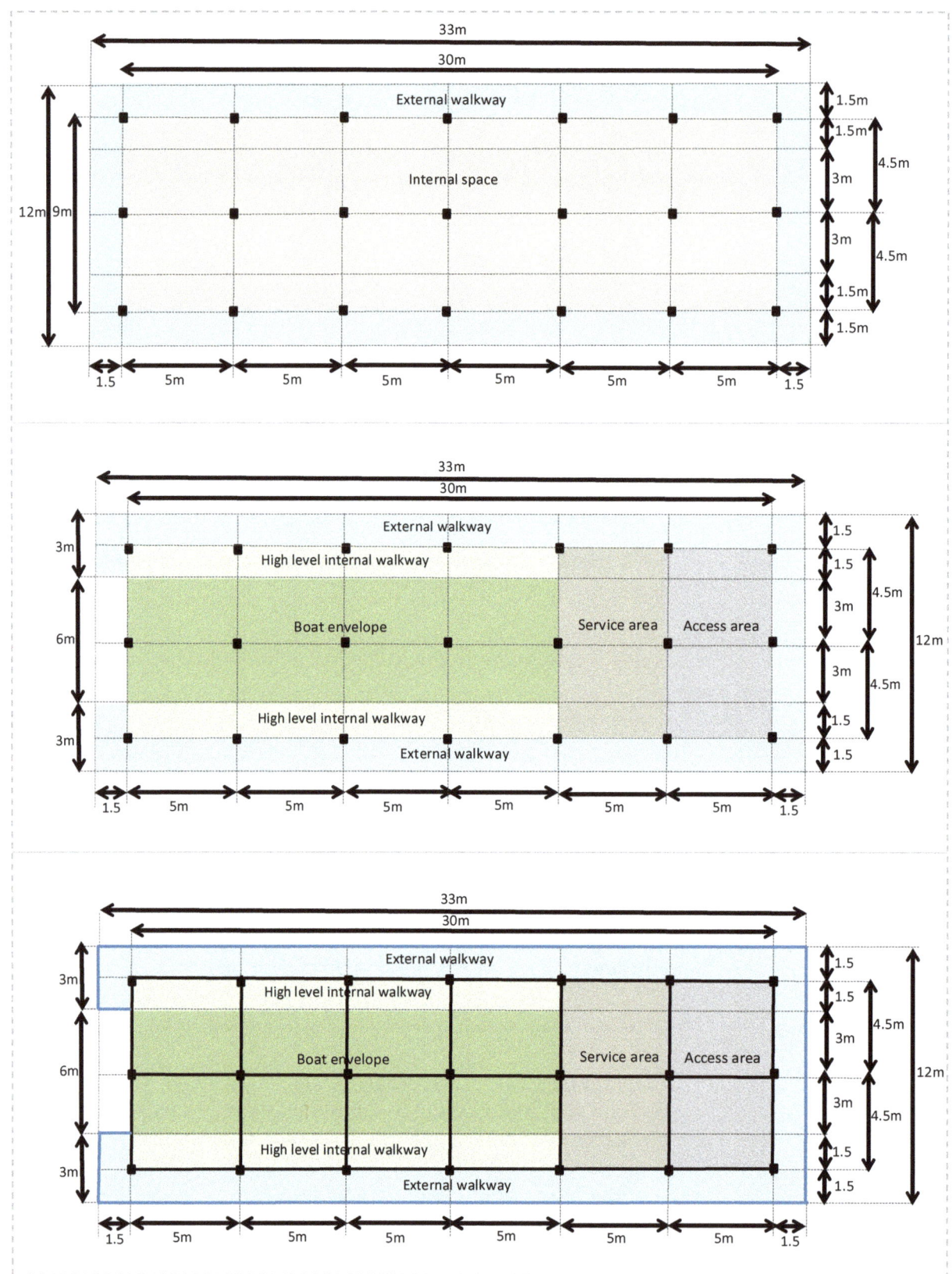

Figure 34: Designs of the framing plans

HIGH-LEVEL WALKWAY	A row of horizontal columns is placed on the horizontal grid line at 1.5 from the external envelope, to support the walkway. Beams are placed accordingly.
OFFICE	The column pattern of the high-level walkway level is repeated. Beams are placed to connect all columns. A design decision is made to allow a full-height gallery above the access area to accommodate lift and stair access to the office floor.
ROOF	The column pattern of the high-level walkway level is repeated. Functional areas are differentiated by colour. Beams are placed on the perimeter of the external envelope and the internal functional areas.

Figure 35: Technical details

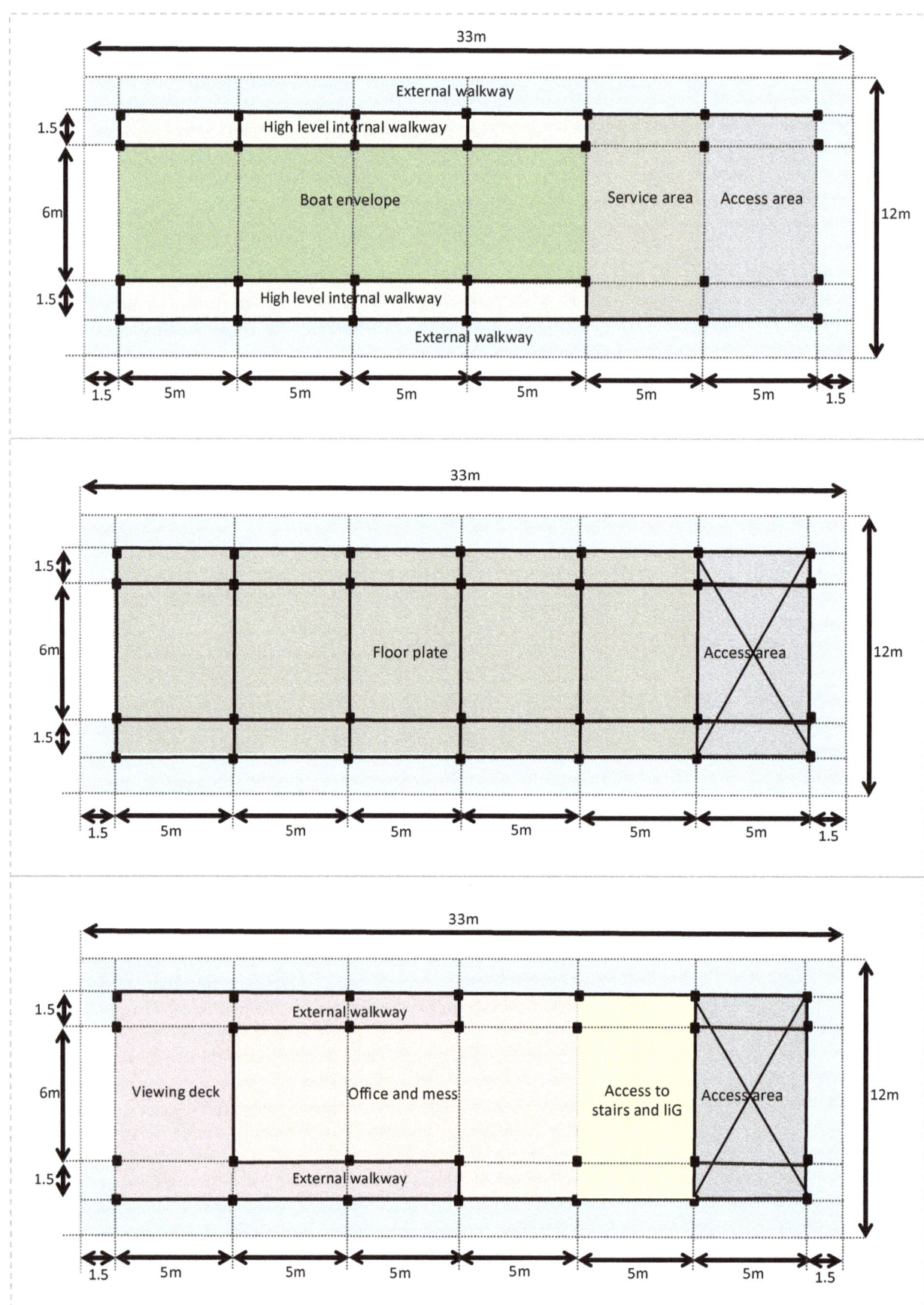

Figure 36: Technical details

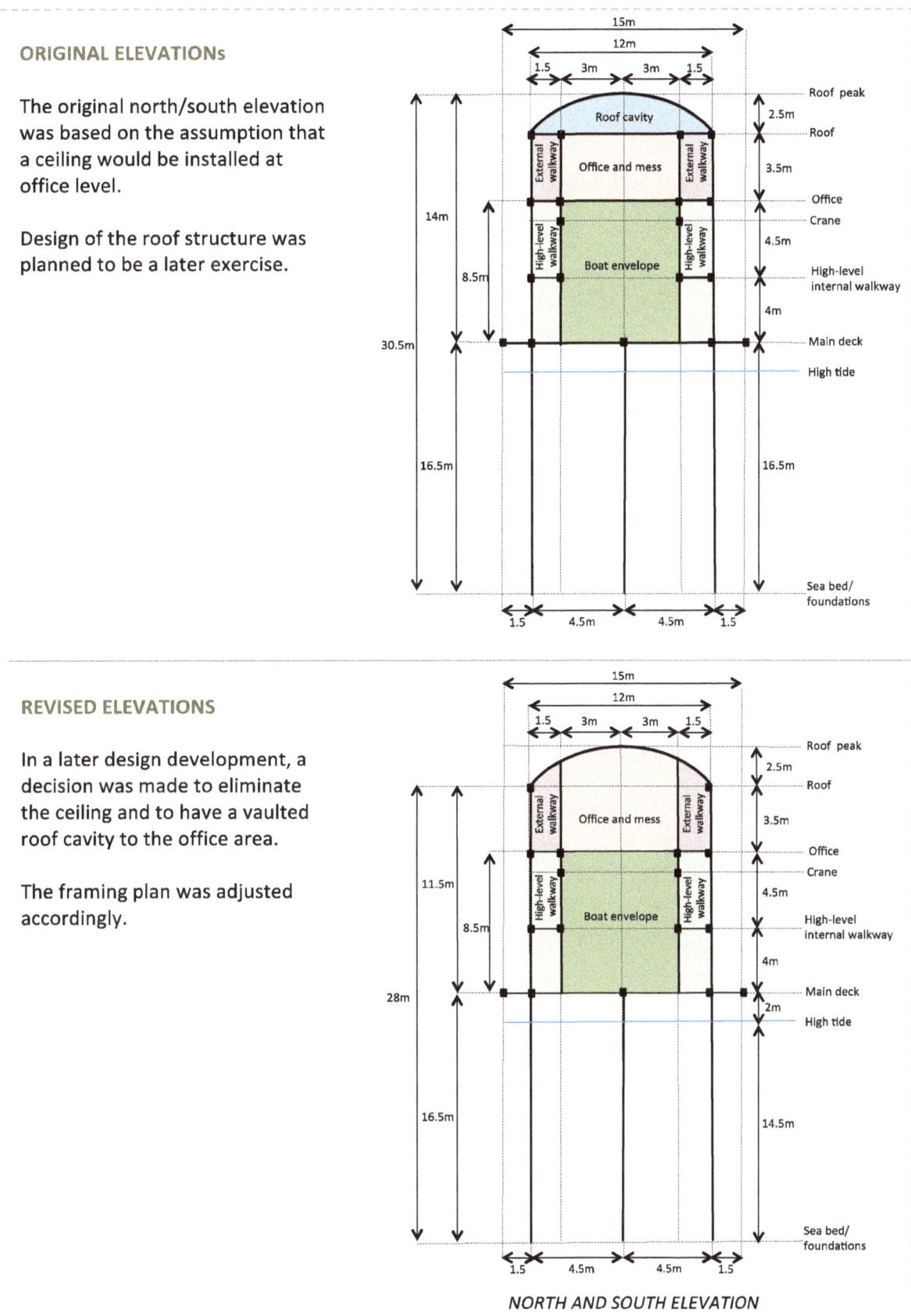

Figure 37: North and south elevation

FRAMING: ELEVATIONS

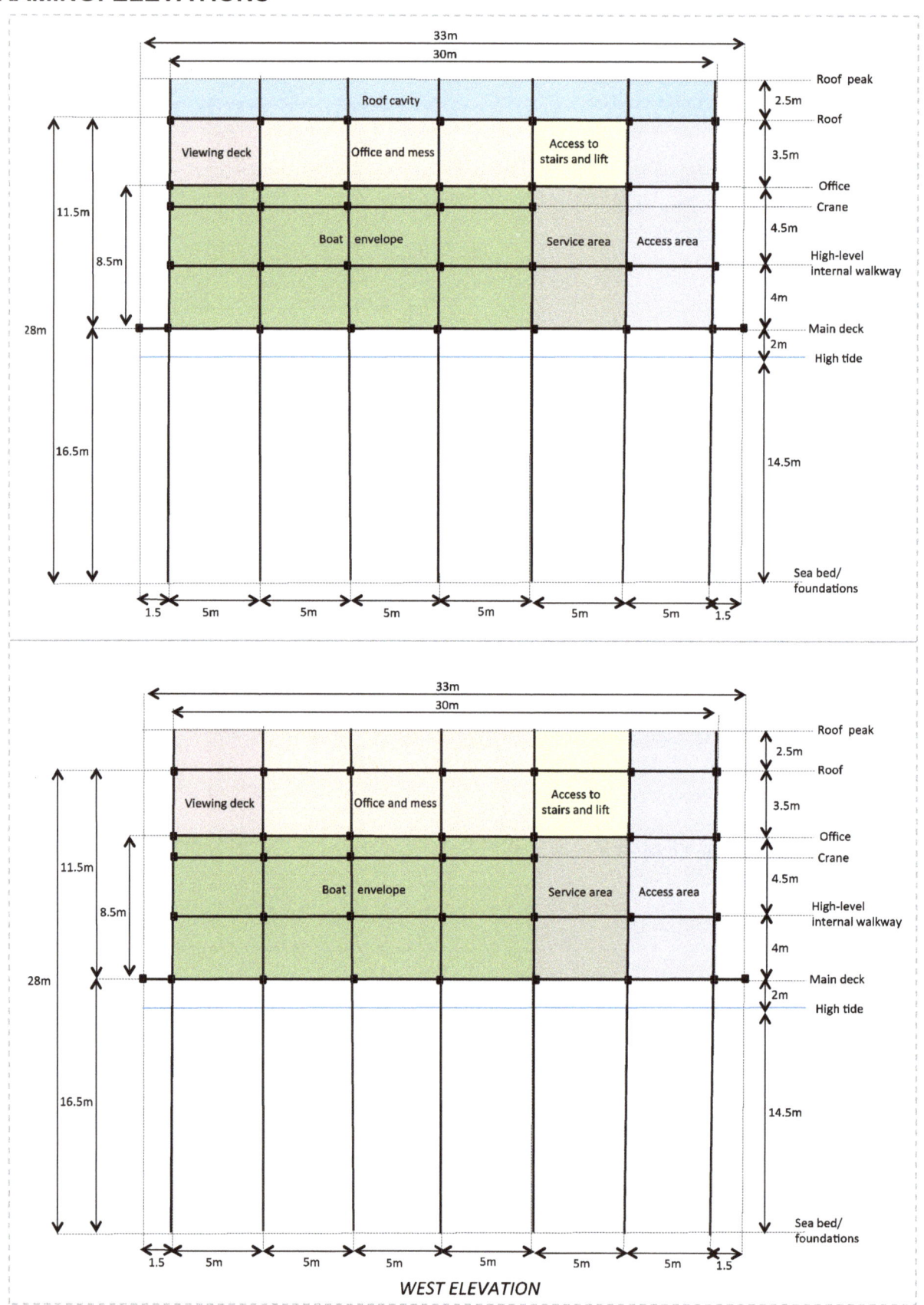

WEST ELEVATION

Figure 38: Framing elevations

The height of the main deck above the foundations is consideraable: 16.5m. Taking buildability and transport into account, it was decided, for premliminary planning purposes, to subdivide this height into evenly distributed 4m components (as shown in the diagrams alongside)

In the Revit model, after adjustments were made for various factors, the length of these components is 4.25m.

Beams are placed, to connect the columns, as shown in maroon in the diagrams alongside.

In the first and last bays of the lateral elevation and in each bay of the cross-sectional elevation, beams are placed on each grid line, to accommodate bracing.

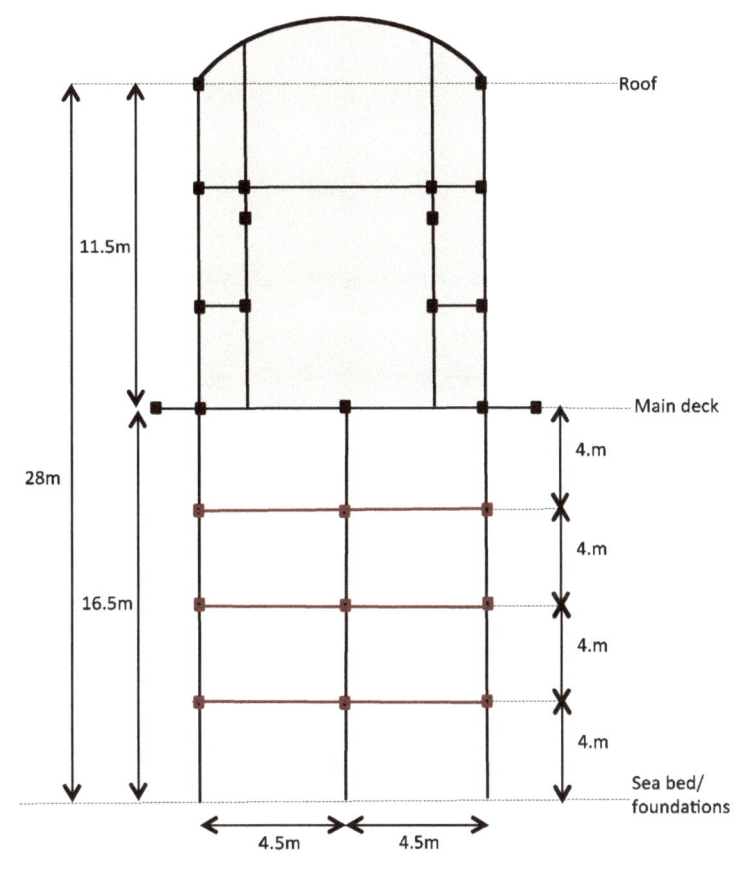

Figure 39: North and south elevation

FRAMING: SUBSTRUCTURE

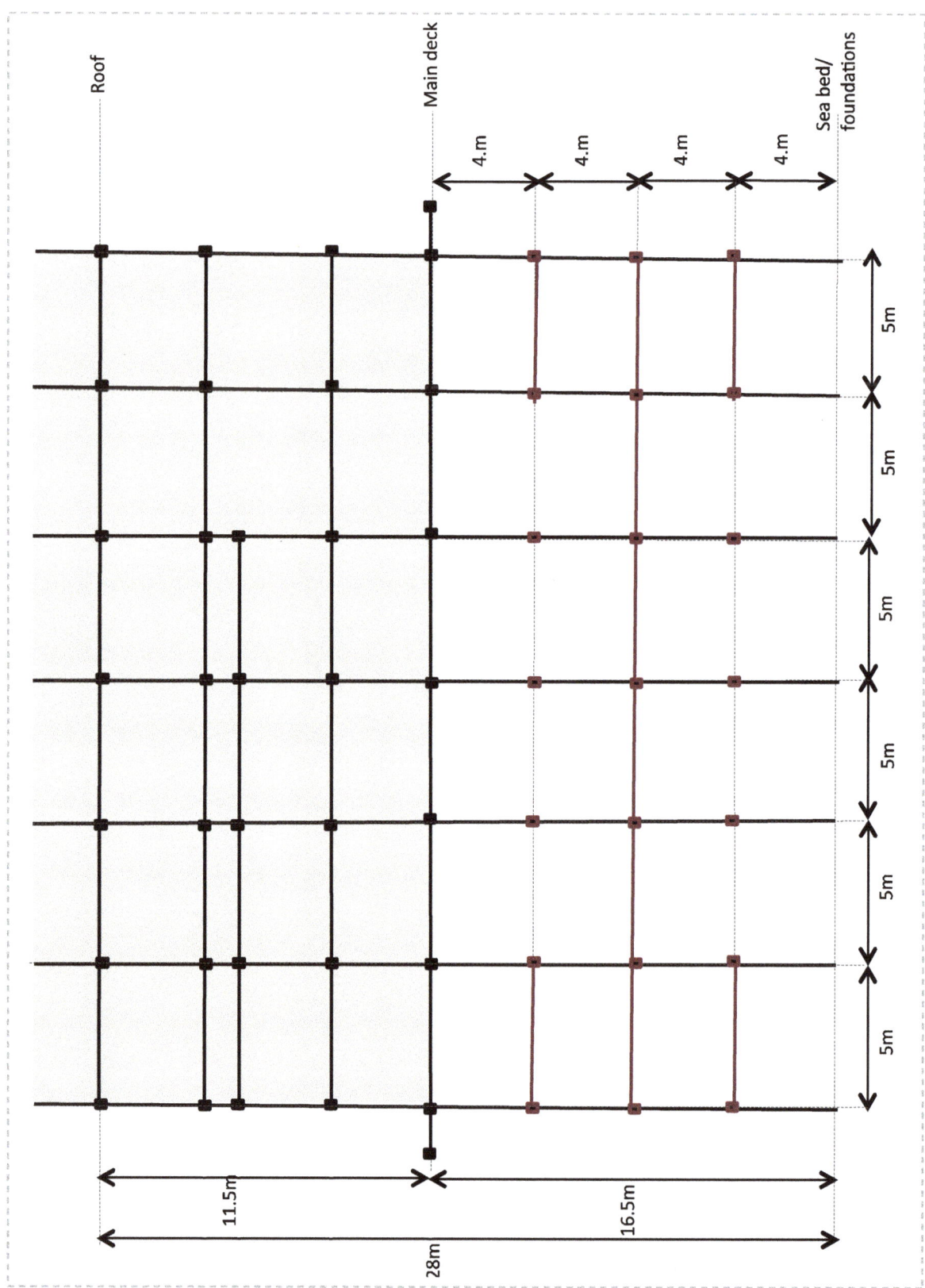

Figure 40: West elevation, framing structure

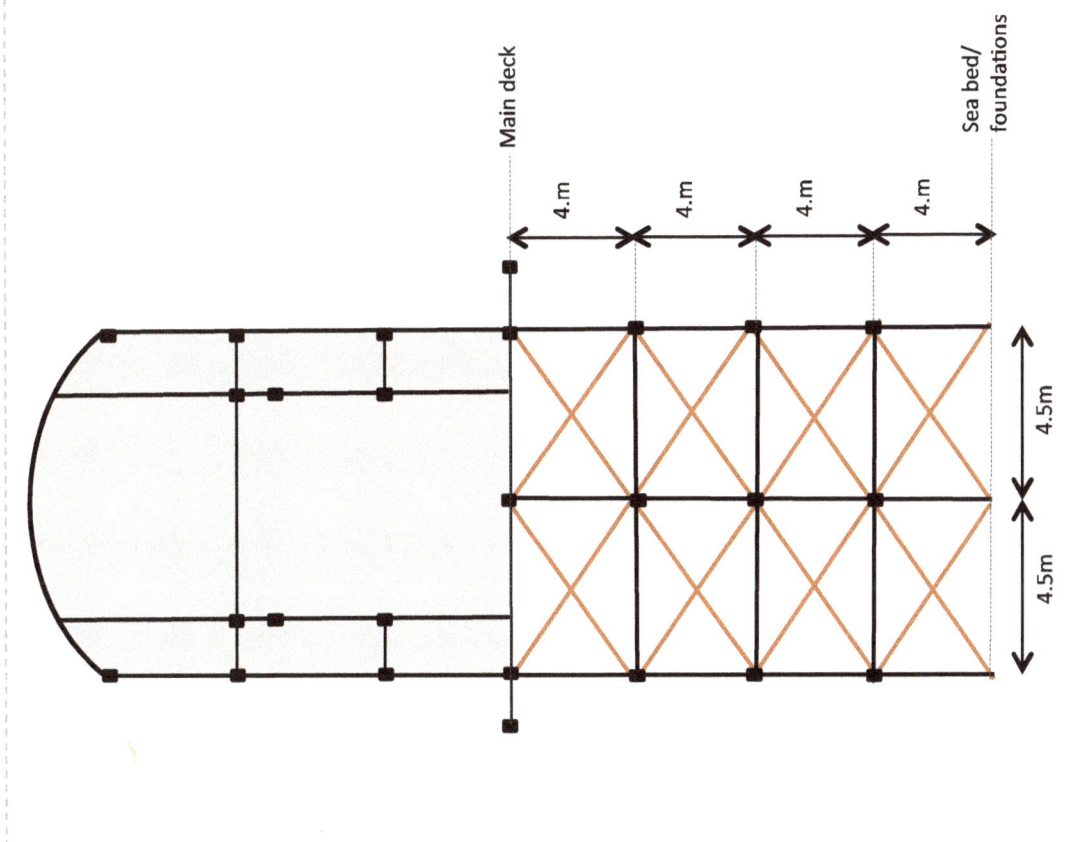

In terms of stability, a design decision was made to employ bracing to provide stability in the north-south direction and portal frame action to provide stability in the west-east direction. Bracing is shown in orange.

In the west and east elevations, bracing is employed in each of the end bays. Bracing is provided in more than one bay, to protect against the risk of failure of the bracing in the other bay.

In the north and west elevations, stability in the superstructure is to be provided by portal frame action. However, in consideration of the considerable forces to be exerted on the substructure by wave action, bracing is to be provided at substructural level of this elevation too, as shown in the diagram alongside.

The issue of stability is discussed in greater detail in a later section.

Figure 41: North and south elevation

BRACING

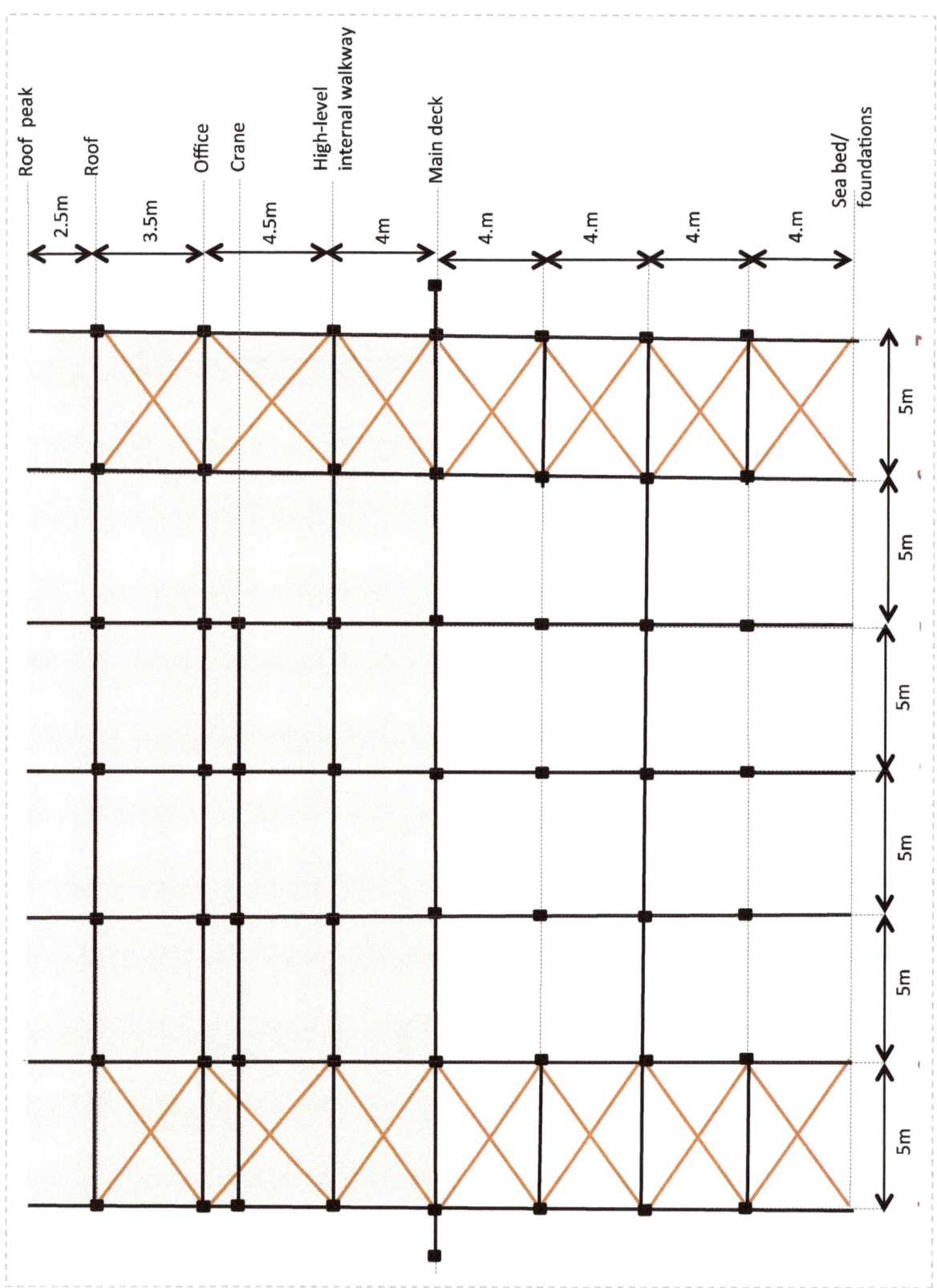

Figure 42: West elevation

LOADING SCHEDULE

ID	Item	Type	Units	Variation	kN/m^2	kg/m	kg/m^2	X(m)	Y(m)	Z(m)	Area (m^2)	Length (m)	Force (kN)	Weight (kg)	Support
	Roof level														
1	Roof paneling	Dead	1	Steel thickness (115mm)			12.50	9.9	30.4				37.0	3,771.5	7
2	Cellular roof beams	Dead	7	Cellular (796x210x82)	-	82.00						11.2	63.3	6,450.6	7
3	Purlins	Dead	3	Sleeved single Span (10m) 300 Z 30		11.97		30.4					10.7	1,092.6	7
4	Services	Dead	1		0.80			0.3	30.4				7.3	744.3	7
5	Snow	Live	1		0.60			0.5	30.4				9.1	930.4	7
	Office floor level														
6	Beams	Dead	24	UB305x102x25		25.00						5.0	29.4	3,000.0	7
7	Columns	Dead	14	UB305x102x25		25.00						3.3	11.2	1,137.5	15
8	Columns	Dead	8	UB305x102x25		25.00						4.5	8.8	900.0	15
9	Bracing	Dead	4	SHS 60x60		5.30						6.0	1.2	127.2	15
10	Curtain wall (internal)	Dead	3		0.40						31.8		38.1	3,888.7	15
11	Curtain wall (external)	Dead	3		0.60						61.8		111.2	11,339.4	15
12	Internal partitions	Dead	9		0.50						11.0		49.5	5,045.9	15
13	Architectural details	Dead	20		0.05						10.0		10.0	1,019.4	15
14	Stairs	Dead	1		4.79			4.8		2.7			61.5	6,271.7	15
15	Floor	Dead	1		2.36						245.0		578.2	58,939.9	
	High level walk way														
16	Curtain wall (external)	Dead	3		0.60						123.6		222.5	22,678.9	18
17	Stairs	Dead	1		4.79			4.8		2.7			61.5	6,271.7	18
18	Architectural details	Dead	10		0.05						10.0		5.0	509.7	18
18	Floor	Dead	1		2.36						76.3		180.0	18,353.1	
	Main deck level														
19	Curtain wall (external)	Dead	3		0.60						123.6		222.5	22,678.9	30
20	Internal partitions	Dead	4		0.50						72.0		144.0	14,678.9	30
21	Beams	Dead	44	UB305x102x25		25.00						5.0	54.0	5,500.0	30
22	Columns	Dead	28	UB305x102x25		25.00						8.5	58.4	5,950.0	30
23	Crane	Dead	1										100.0		30
24	Crane rail (CR100)	Dead	1			89.00						50.0	43.7	4,450.0	30
25	Crane operational weight	Live	1										19.6	2,000.0	30
26	Bracing	Dead	16	SHS 60x60		5.30						6.0	5.0	508.8	30
27	Curtain wall (external)	Dead	1		0.60						82.4		49.4	5,037.3	30
28	Shutter	Dead	1		0.30						49.0		14.7	1,498.5	30
29	Beams (walkway)	Dead	7	UB305x102x25		25.00						5.0	8.6	875.0	30
30	Architectural details	Dead	18		0.05						10.0		9.0	917.4	30
31	Floor (whinch)	Dead	2		2.36						11.0		51.9	5,292.6	30
32	Ramp (internal)	Dead	1		1.20						11.0		13.2	1,345.6	30
29	Floor (whinch deck)	Dead	1		1.20						23.0		27.6	2,813.5	30
30	Deck	Dead	2		0.40						102.0		81.6	8,318.0	30
31	Whinch	Live	1										3.9	400.0	30
30	Floor	Dead	1		2.36						304.0		717.4	73,133.5	
	Ground foundation														
31	Beams	Dead	100	UB305x102x25		25.00						5.0	122.6	12,500.0	33
32	Bracing	Dead	64	SHS 60x60		5.30						6.0	20.0	2,035.2	33
33	Columns	Dead	21	UB305x102x25		25.00						17.6	90.6	9,240.0	34
34	Foundations	Dead	21										60.0	6,116.2	

Figure 43: Loading schedule

MODULAR COMPONENTS

Technical Properties of the D225 Series
Data sheet PD — July 2009

Earls Colne Business Park, Earls Colne,
Colchester, Essex, CO6 2NS
Tel: 01787 223931
Email: design@milbank.co.uk
Email: estimating@milbank.co.uk

Section Properties
- Area: 27,938 mm²
- Nab: 99.758 mm
- Inertia: 111,502,265 mm⁴
- Zt: 890,292 mm³
- Zb: 1,117,732 mm³

Material Properties
- Fcu: 53 N/mm²
- Fci: 30 N/mm²
- Fct: 3.5 N/mm²
- Et: 29.5 kN/mm²
- Ew: 36 kN/mm²
- Es: 200 kN/mm²

Wire 5mm & 7mm stabilised to B.S. 5896, 1980.
Cement to B.S. 12. One hour fire resistance.

Type D225/5 (1No 5mm + 4No 7mm Wires)

Ms = 15.10 kNm
Mu = 26.11 kNm
Vco = 34.04 kN

Dimensioned Section (225 × 155, with 100 top width, 125 depth detail)

Block Type	Block Density	Self Weight kN/m² (Joist, Blocks & Grout)		
		Single Beam	Double Beams	Triple Beams
Light	650 kg/m³	S540...1.75 S428...2.04 S315...2.55	D695...2.54 D583...2.90 D470...3.45	T850...3.04 T738...3.40 T625...3.90
Medium	1450 kg/m³	S540...2.38 S428...2.63 S315...3.06	D695...2.98 D583...3.29 D470...3.72	T850...3.36 T738...3.66 T625...4.06
Dense	1900 kg/m³	S540...2.75 S428...2.98 S315...3.38	D695...3.31 D583...3.59 D470...4.01	T850...3.67 T738...3.94 T625...4.32

BEAM AND BLOCK FLOORING: MAIN DECK

Technical Properties of the T150 Series
Data sheet PT — July 2009

Earls Colne Business Park, Earls Colne,
Colchester, Essex, CO6 2NS
Tel: 01787 223931
Email: design@milbank.co.uk
Email: estimating@milbank.co.uk

Section Properties
- Area: 14,275 mm²
- Nab: 66.696 mm
- Inertia: 27,343,738 mm⁴
- Zt: 328,240 mm³
- Zb: 409,977 mm³

Material Properties
- Fcu: 53 N/mm²
- Fci: 30 N/mm²
- Fct: 3.5 N/mm²
- Et: 29.5 kN/mm²
- Ew: 36 kN/mm²
- Es: 200 kN/mm²

Wire 5mm & 7mm stabilised to B.S. 5896, 1980.
Cement to B.S. 12. Half hour fire resistance.

Type T15/3 (3No 5mm Wires)
Ms = 4.24 kNm
Mu = 6.67 kNm
Vco = 18.77 kN

Type T15/5 (1No 5mm + 2No 7mm Wires)
Ms = 5.94 kNm
Mu = 9.53 kNm
Vco = 19.41 kN

Dimensioned Section (150 × 127, 100 top width, 50 flange)

Block Type	Block Density	Self Weight kN/m² (Joist, Blocks & Grout)		
		Single Beam	Double Beams	Triple Beams
Light	650 kg/m³	S525...1.15 S412...1.33 S300...1.60	D652...1.63 D539...1.84 D427...2.16	T779...1.93 T666...2.15 T554...2.46
Medium	1450 kg/m³	S525...1.84 S412...1.95 S300...2.15	D652...2.15 D539...2.31 D427...2.54	T779...2.36 T666...2.52 T554...2.75
Dense	1900 kg/m³	S525...2.15 S412...2.31 S300...2.48	D652...2.46 D539...2.58 D427...2.77	T779...2.63 T666...2.75 T554...2.94

BEAM AND BLOCK FLOORING: OFFICE FLOOR

DESIGN SHEETS

STRUCTURAL DRAWINGS	S 96	Foundations
	S 97	Substructure A Bracing
	S 98	Substructure B Bracing
	S 99	Substructure C Bracing
	S 100	Main Deck
	S 101	Winch Level
	S 102	High-Level Walkway
	S 103	Crane Level
	S 104	Office Floor
	S 106	Roof
	S 201	North and South Elevations
	S 202	East Elevation
	S 203	West Elevation
	S 300	Structural Sections
	S 301	Structural Sections – Portal Action
	S 501	Levels
ARCHITECTURAL DRAWINGS	A 50	Site
	A 100	Main Deck
	A 101	Winch Level
	A 102	High-Level Walkway
	A 104	Office Floor
	A 201	North and South Elevation
	A 202	East Elevation
	A 203	West Elevation
	A 301	Section 1
	A 302	Section 2
	A 303	Section 3

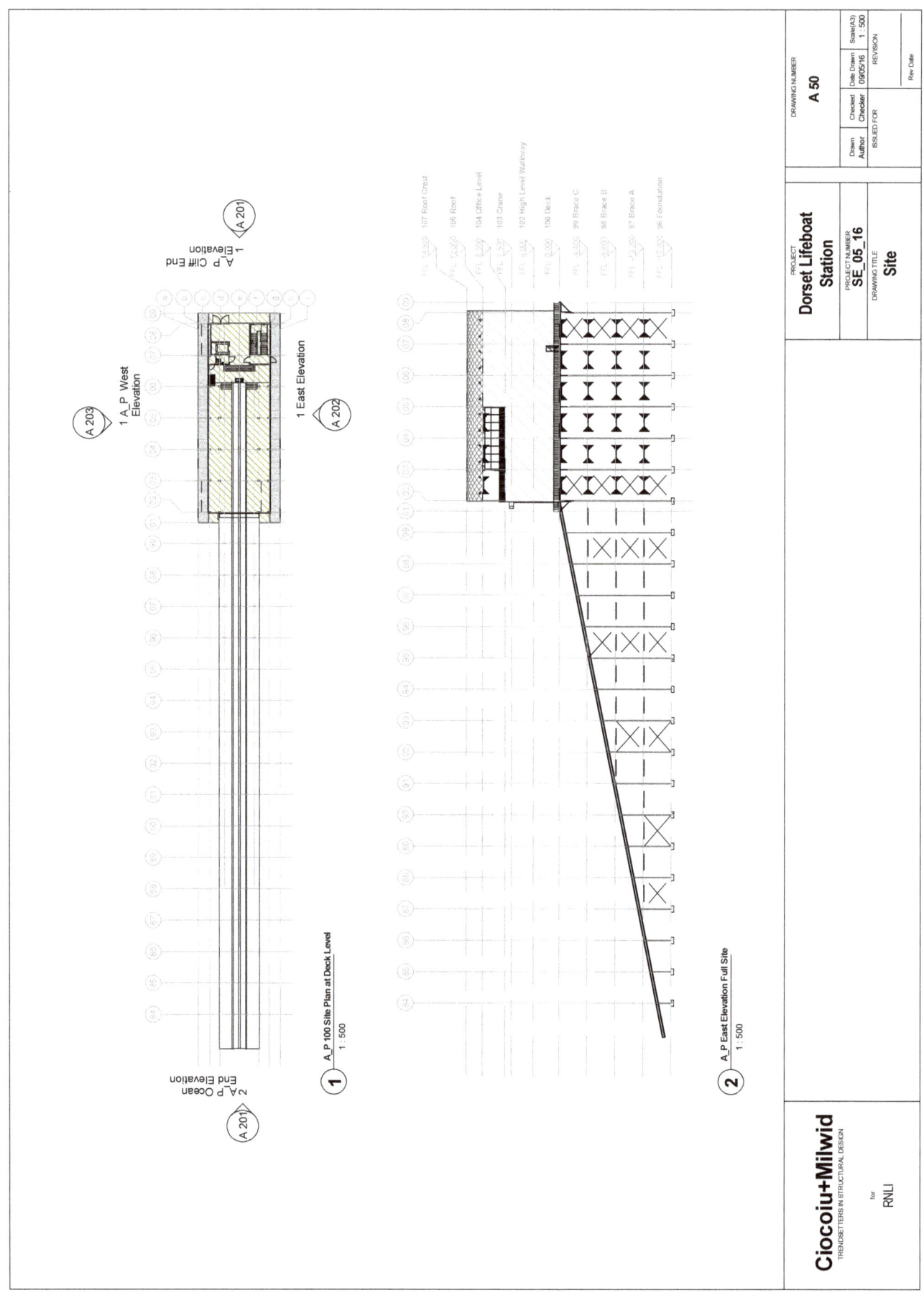

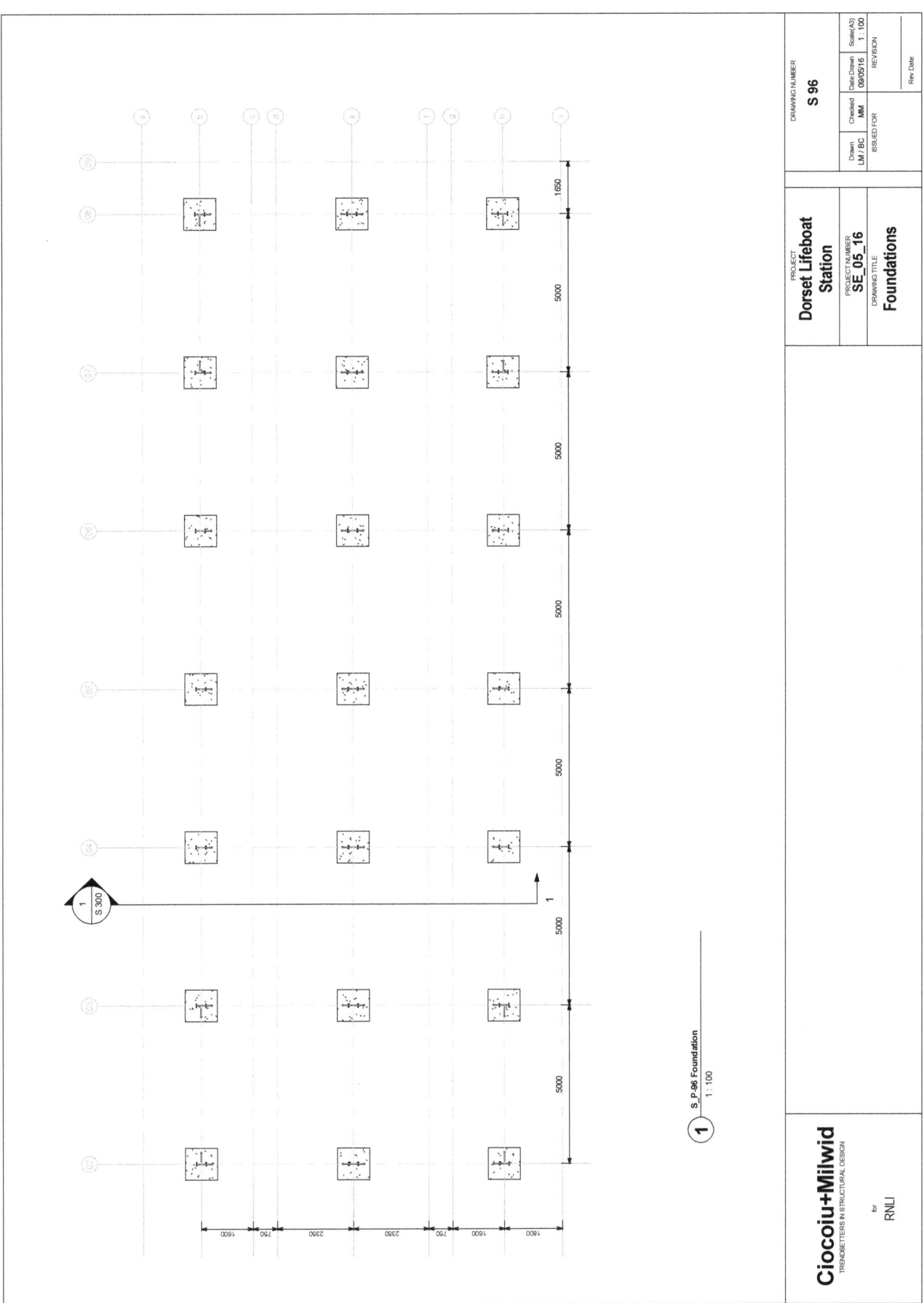

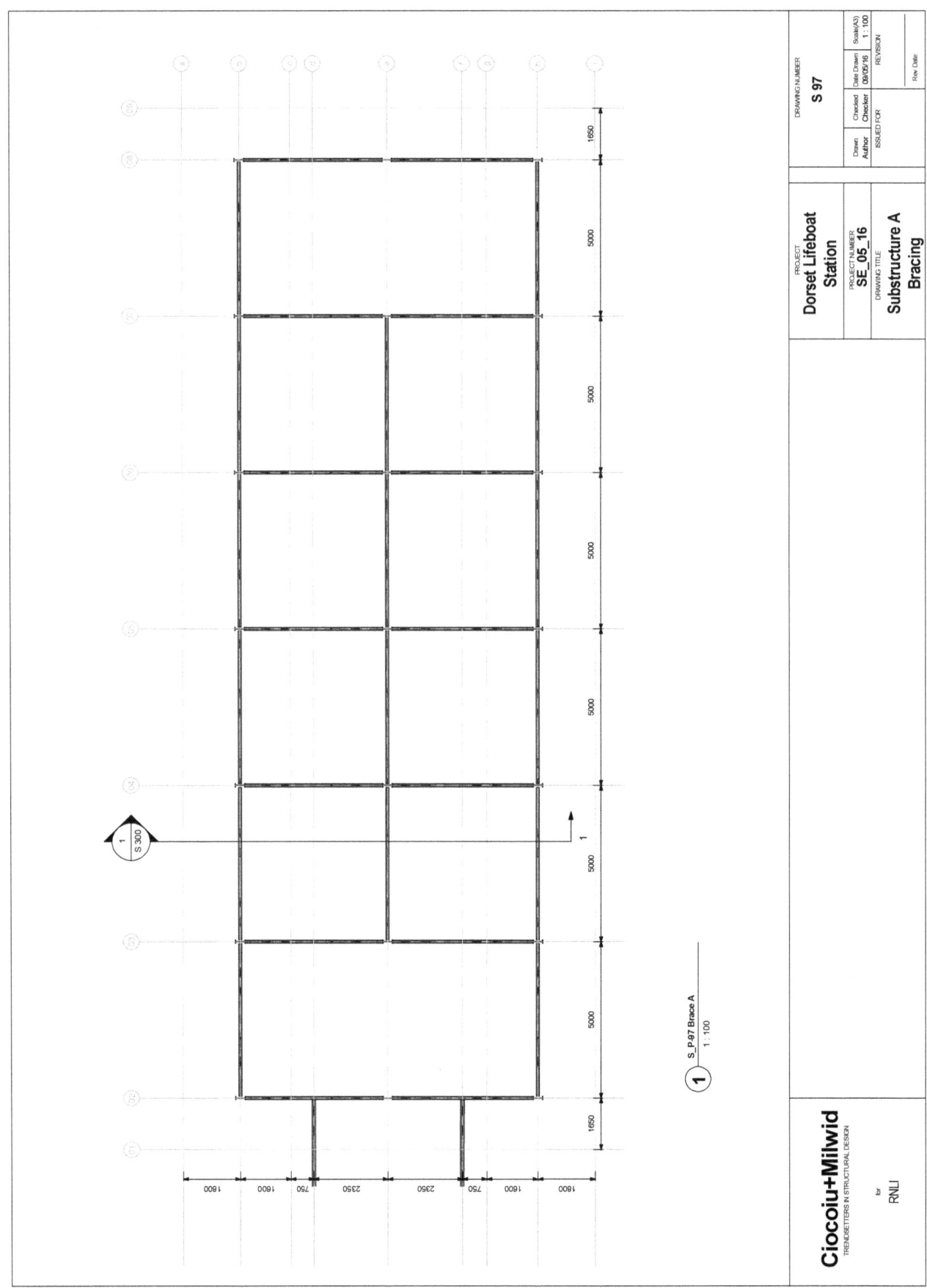

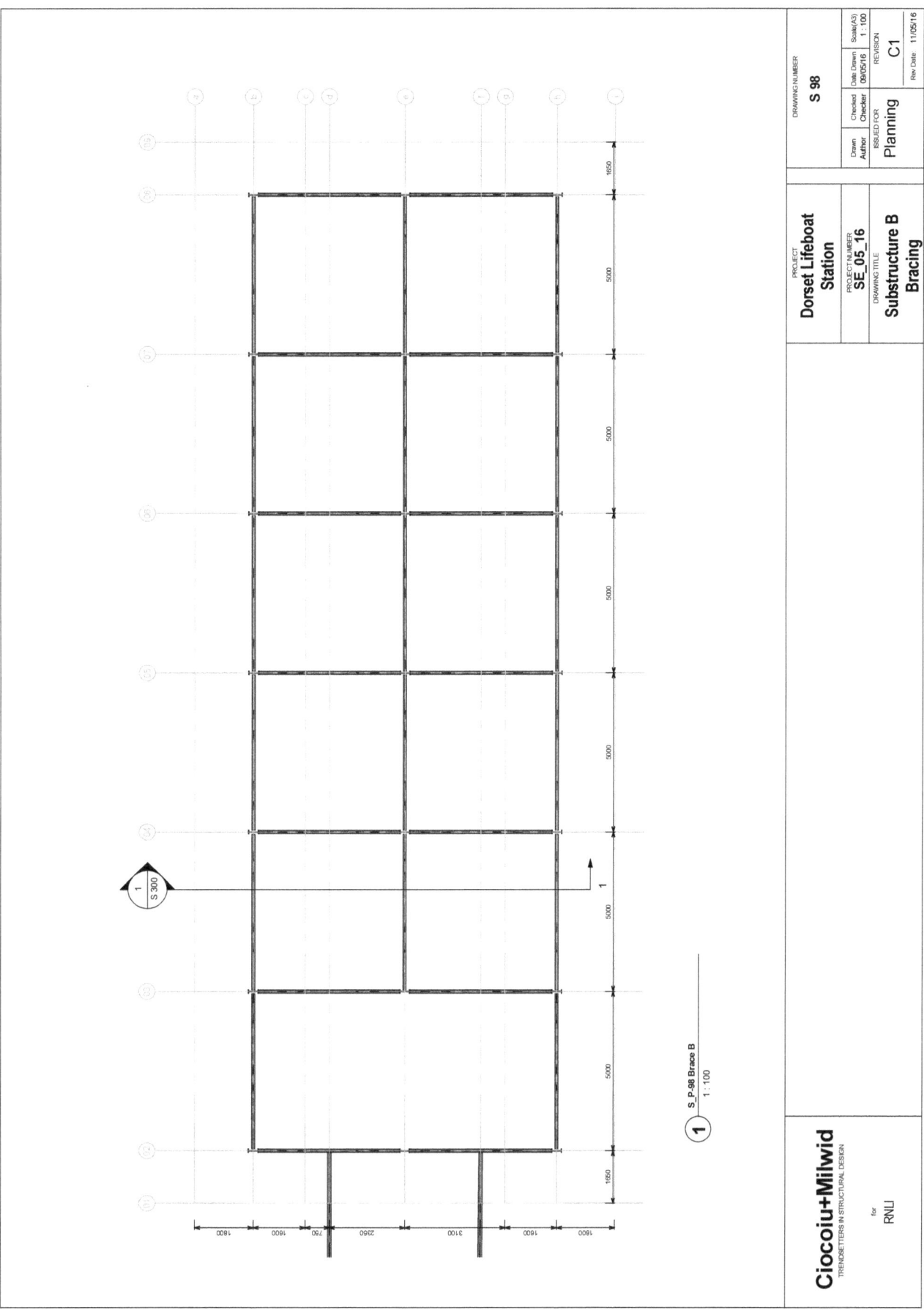

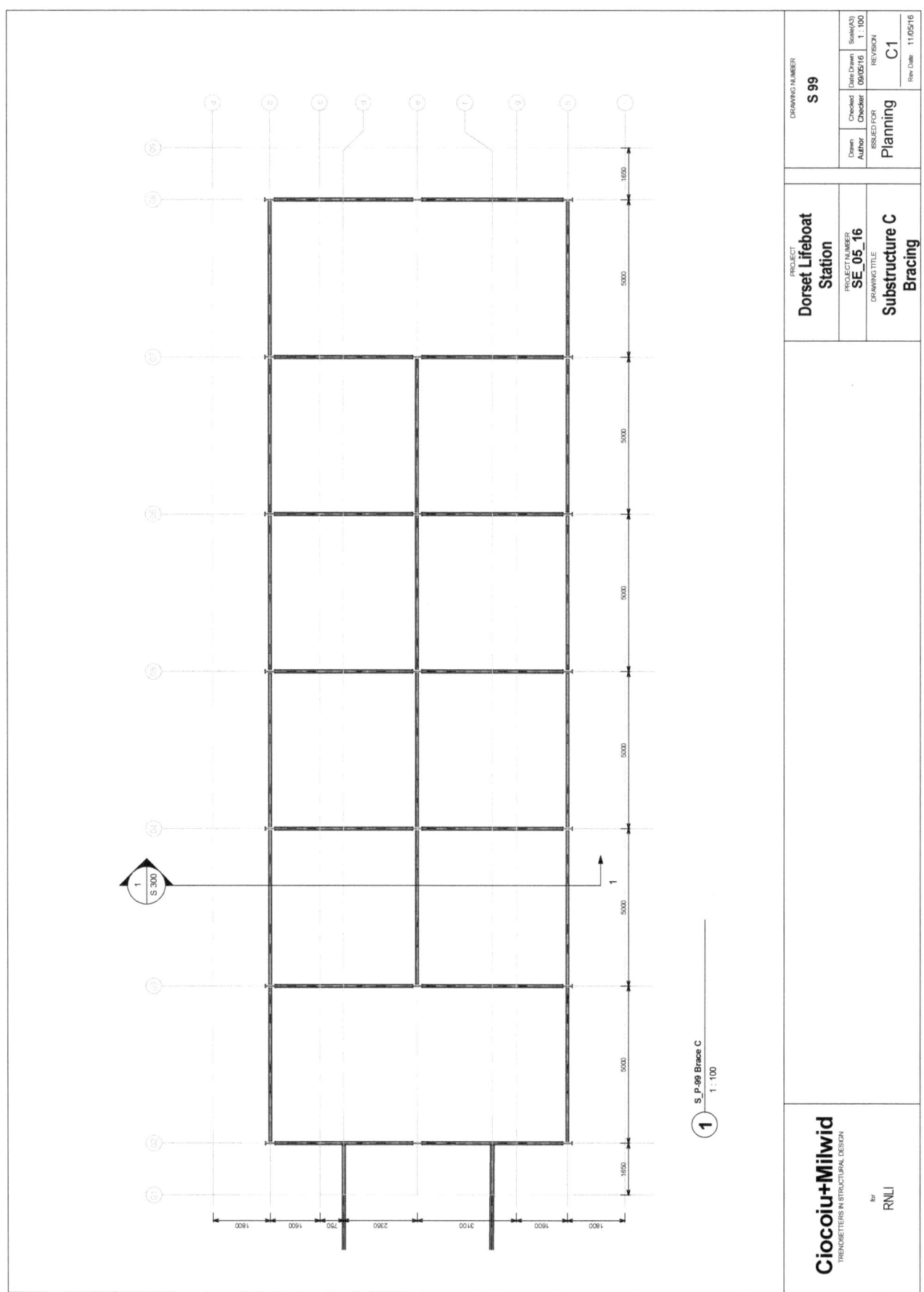

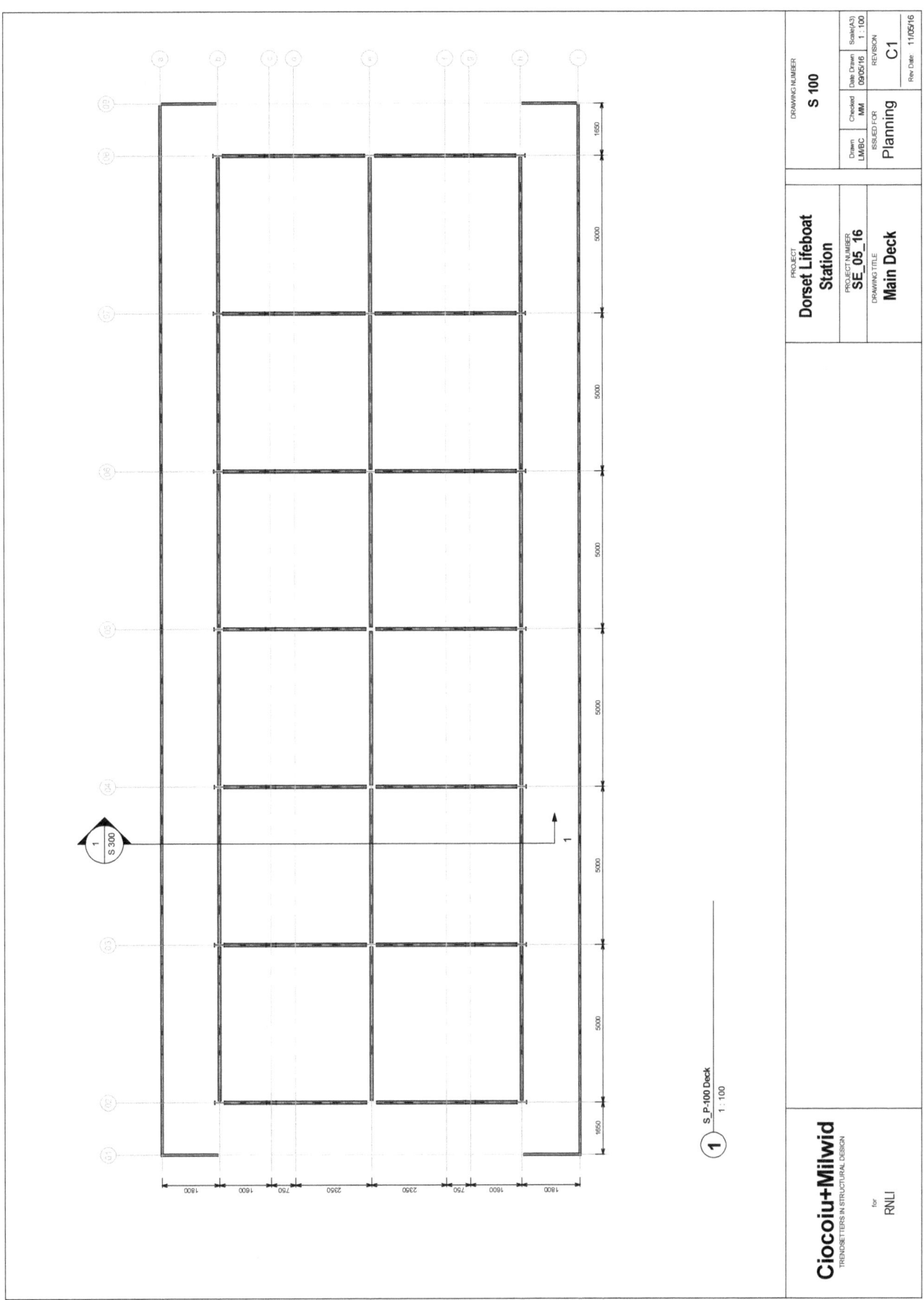

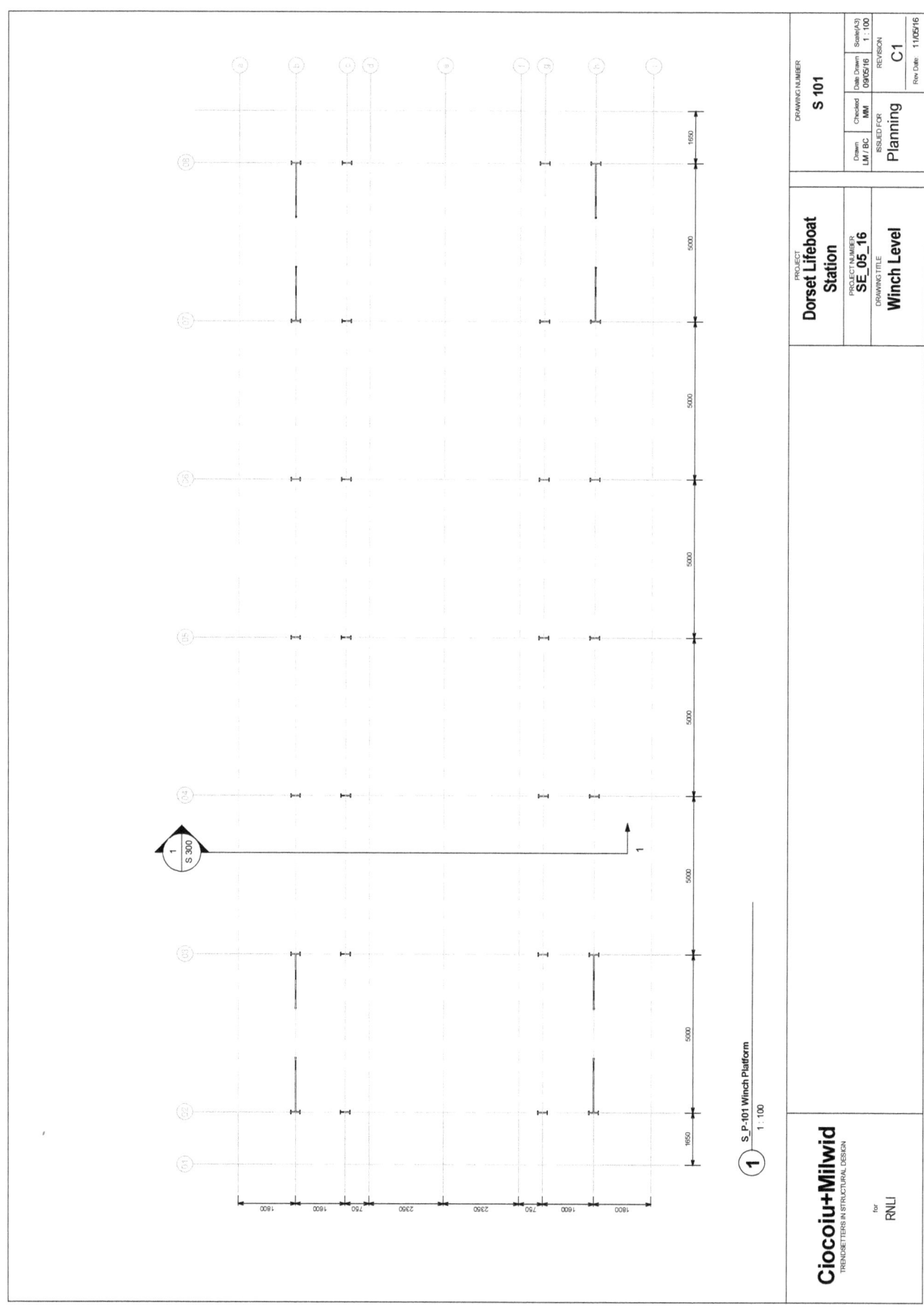

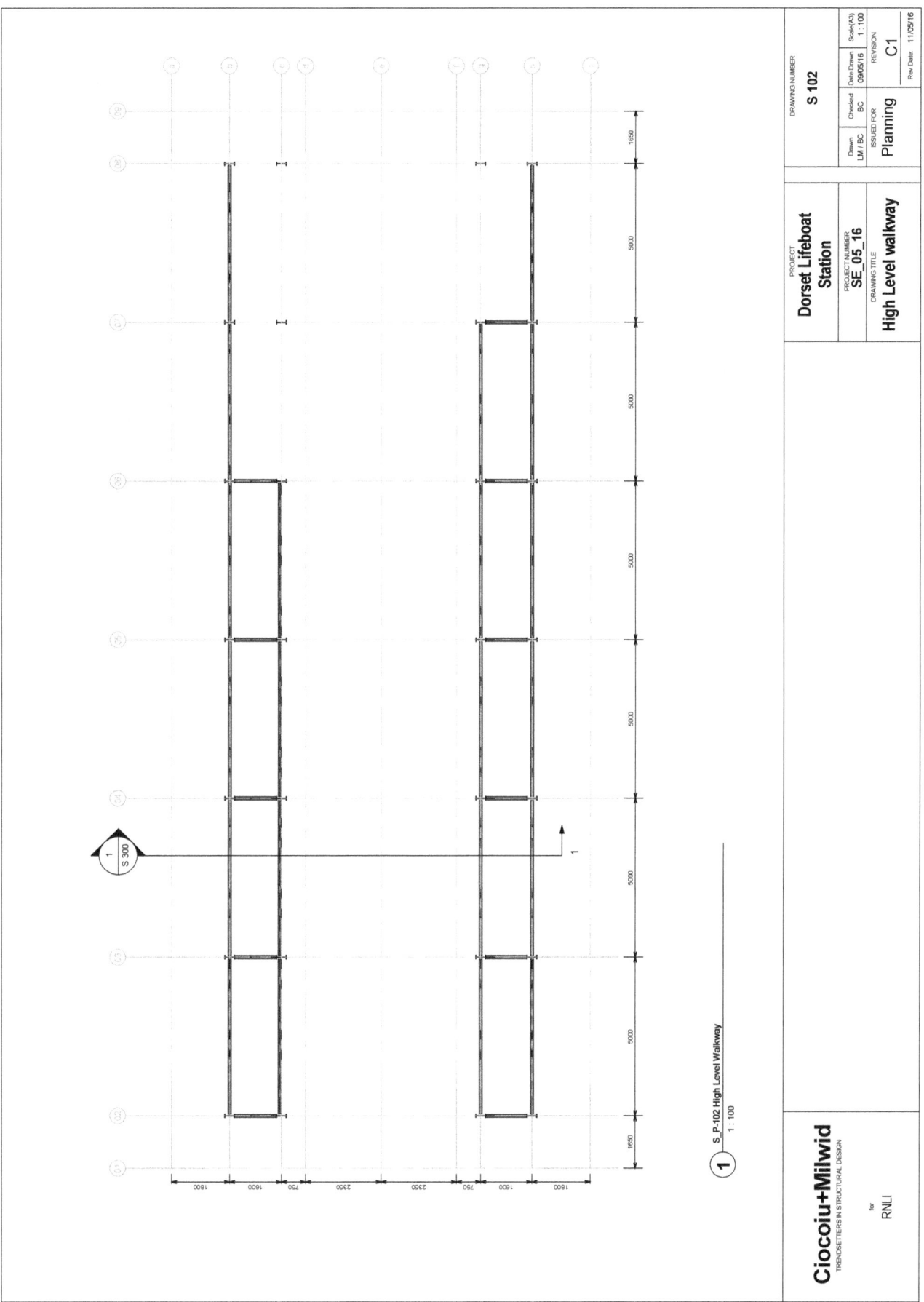

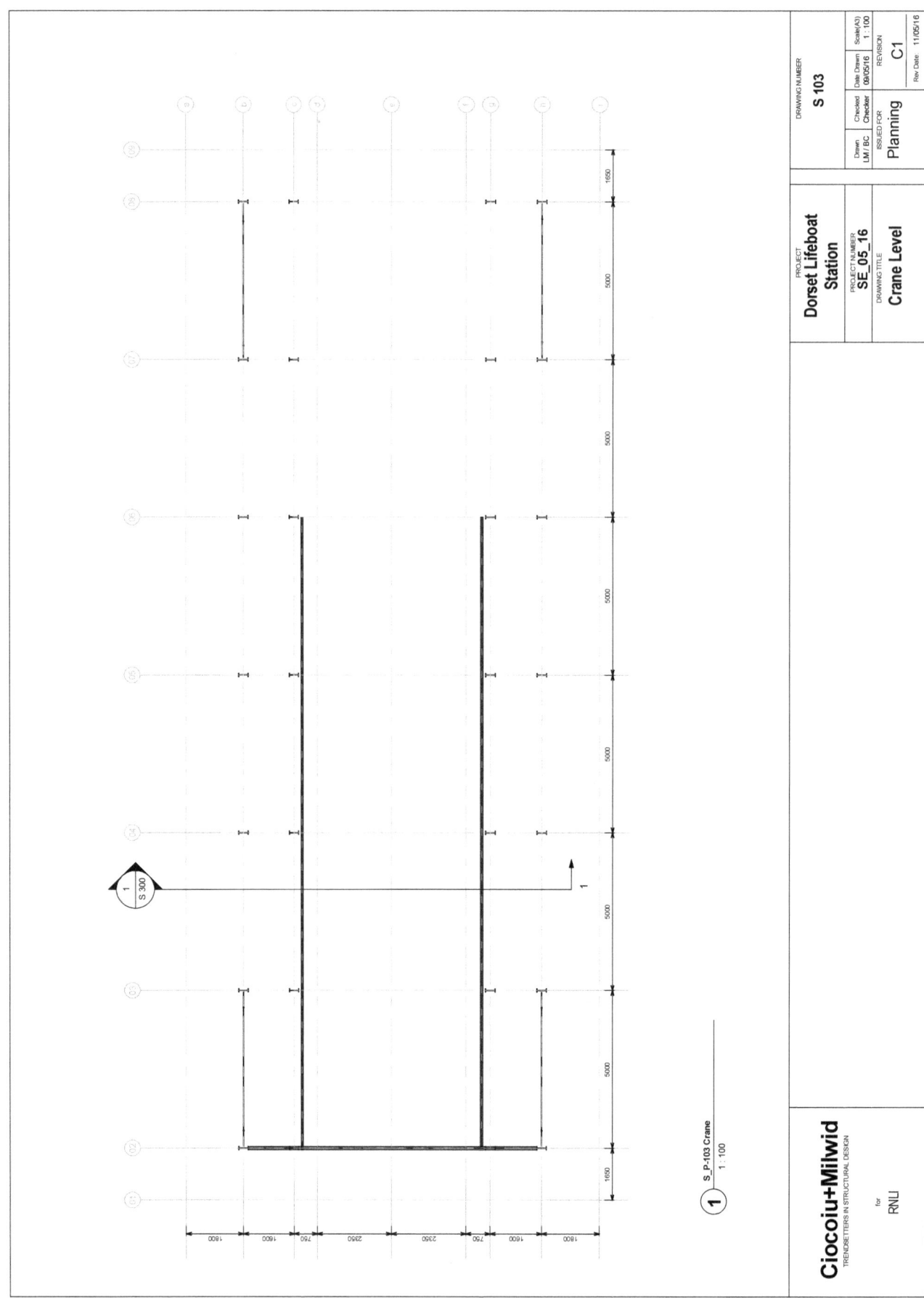

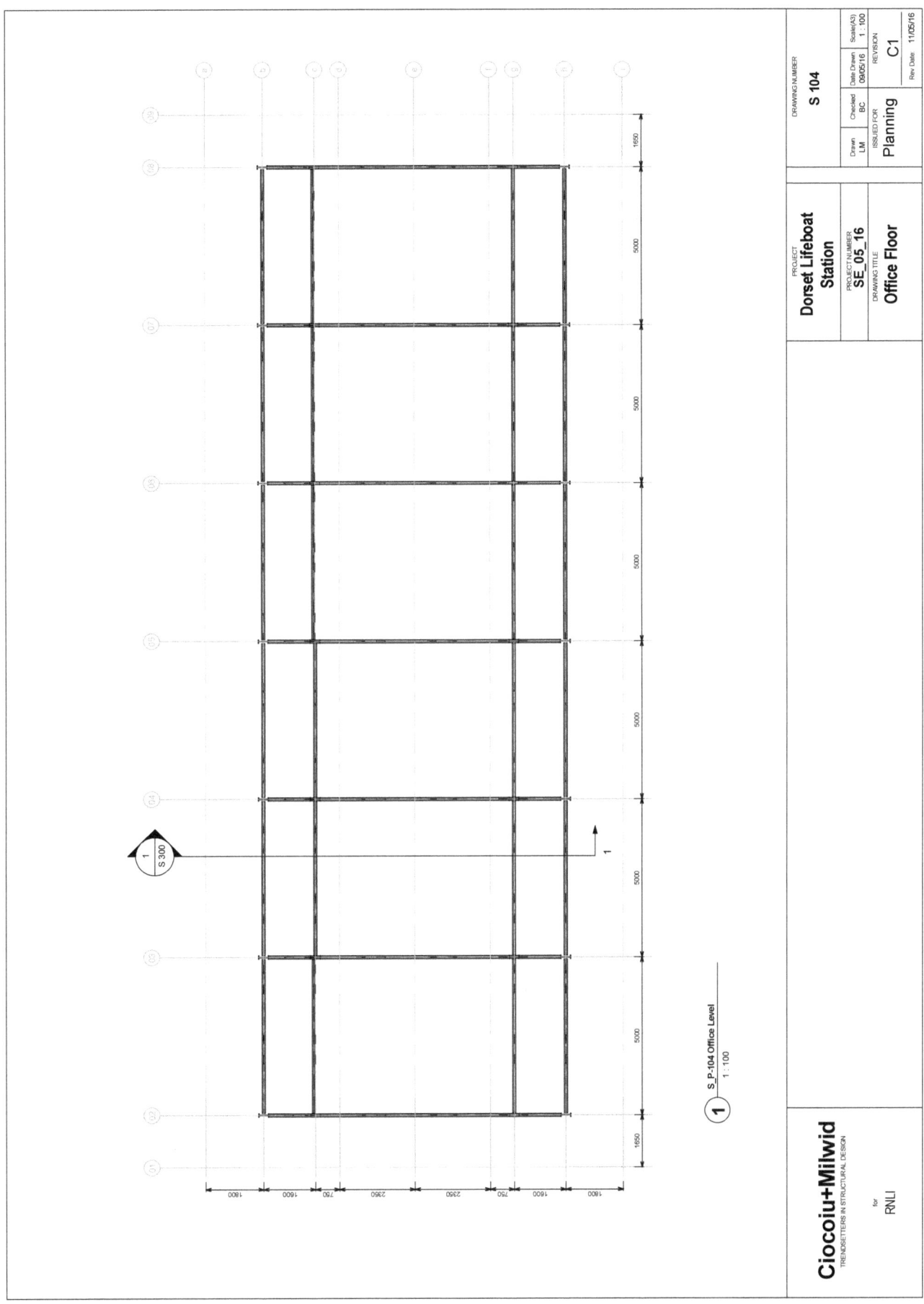

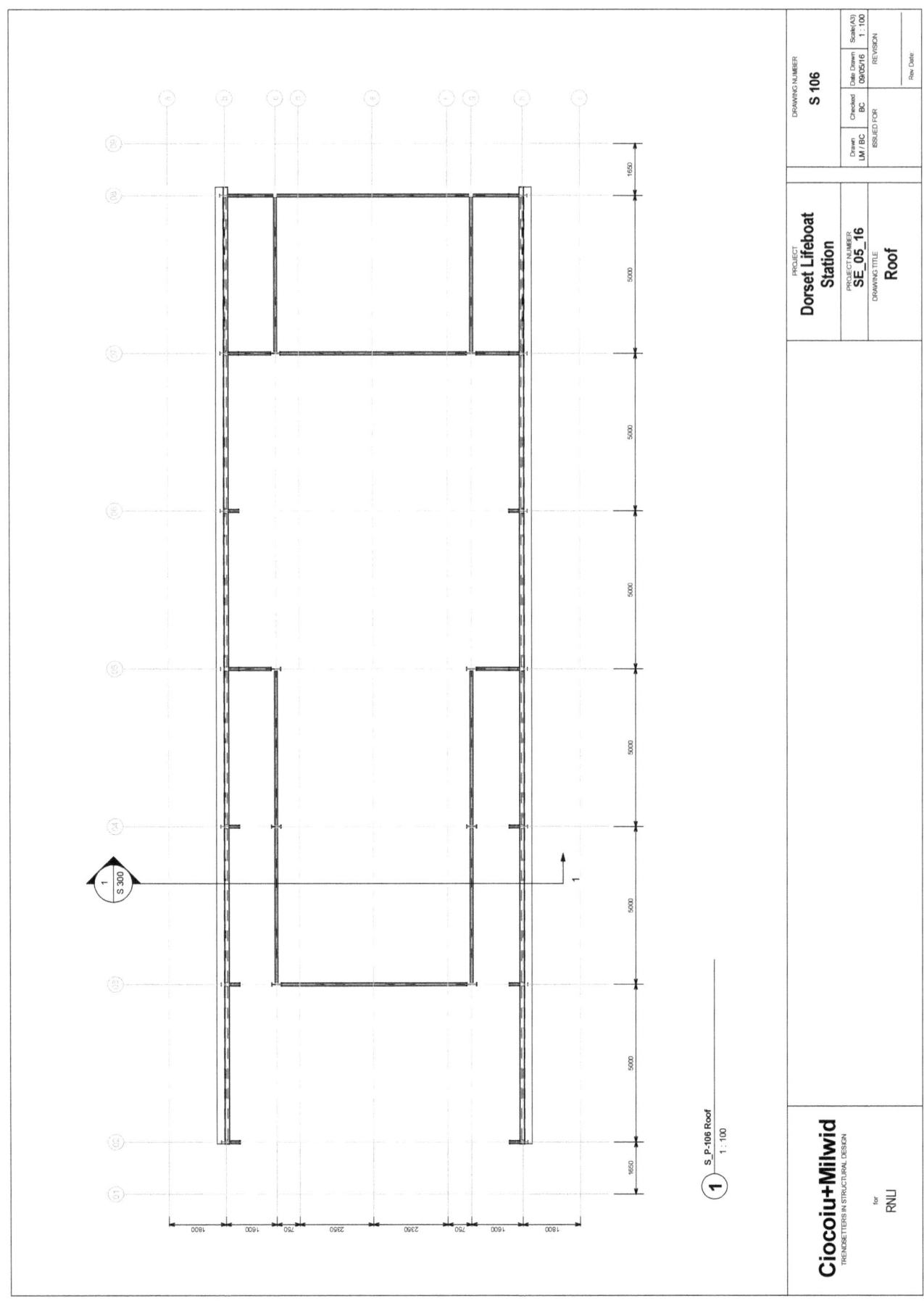

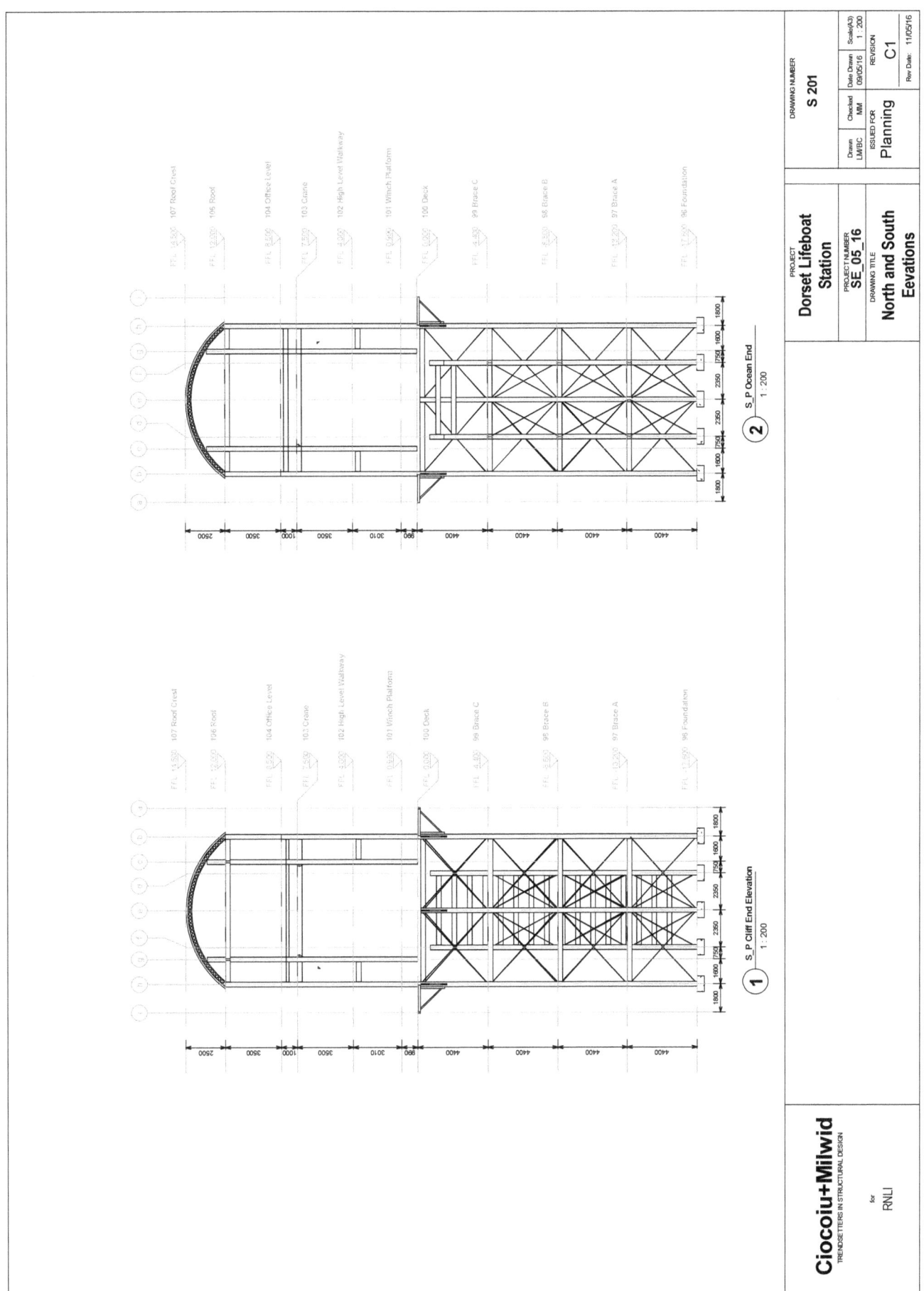

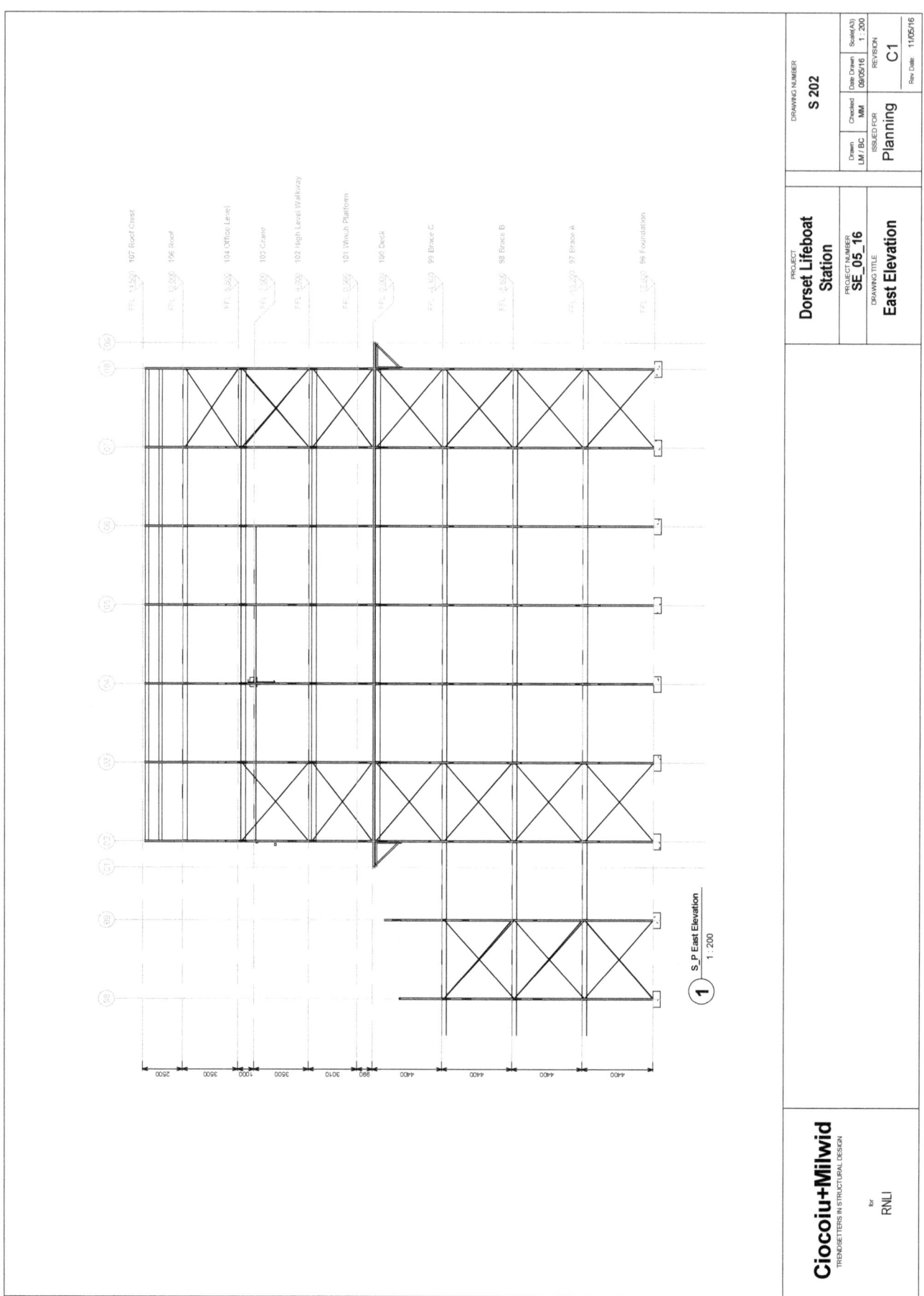

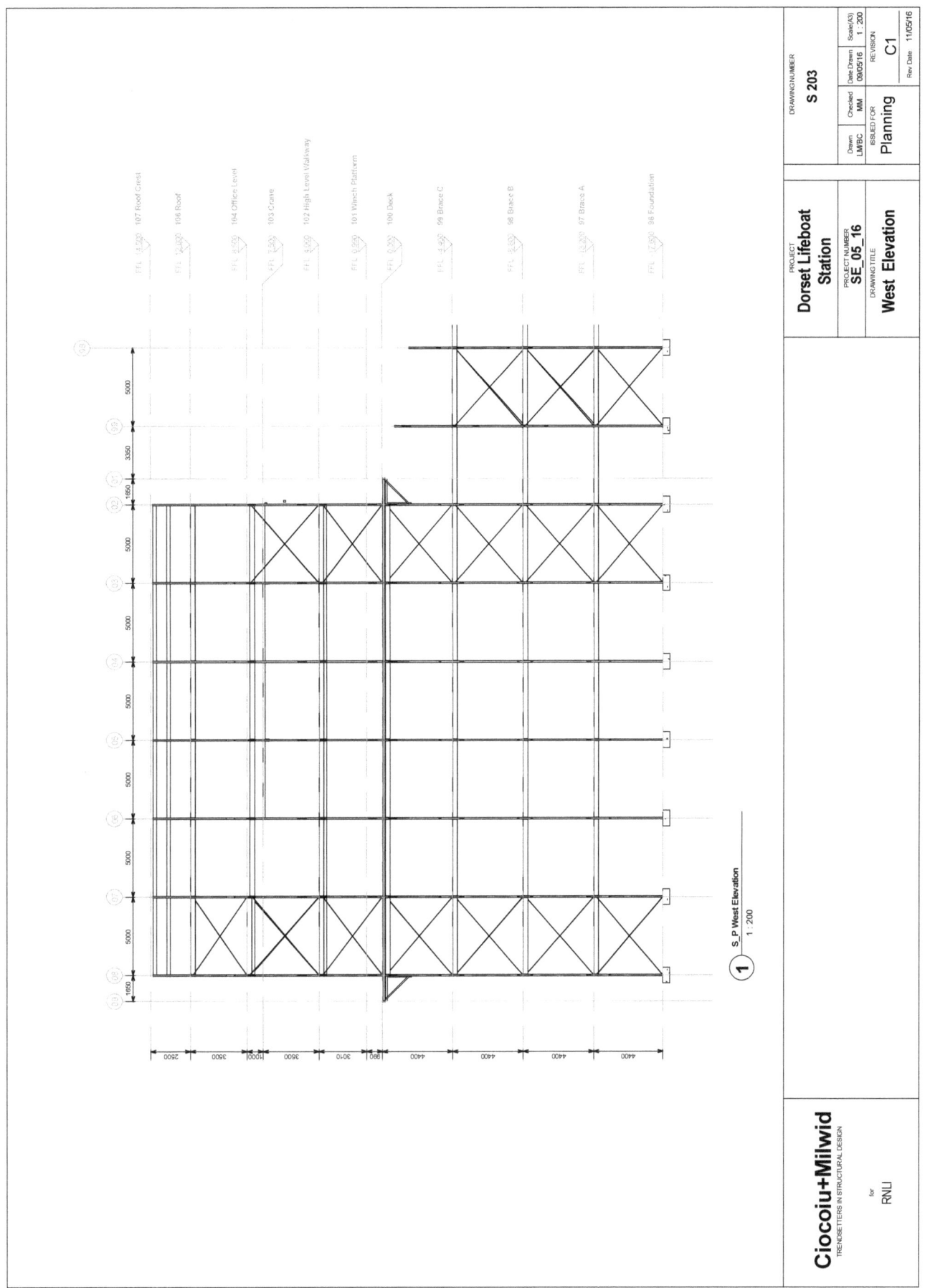

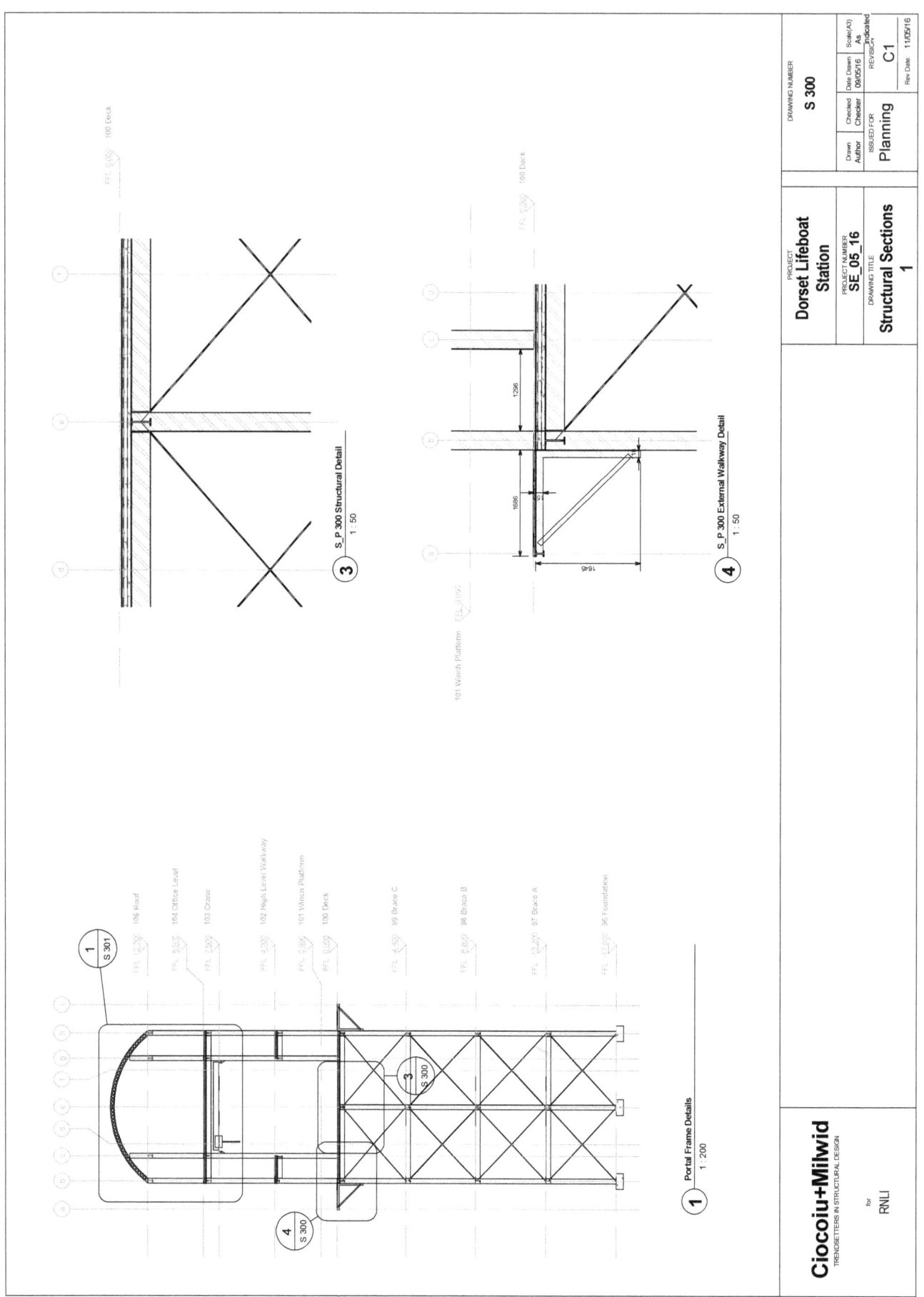

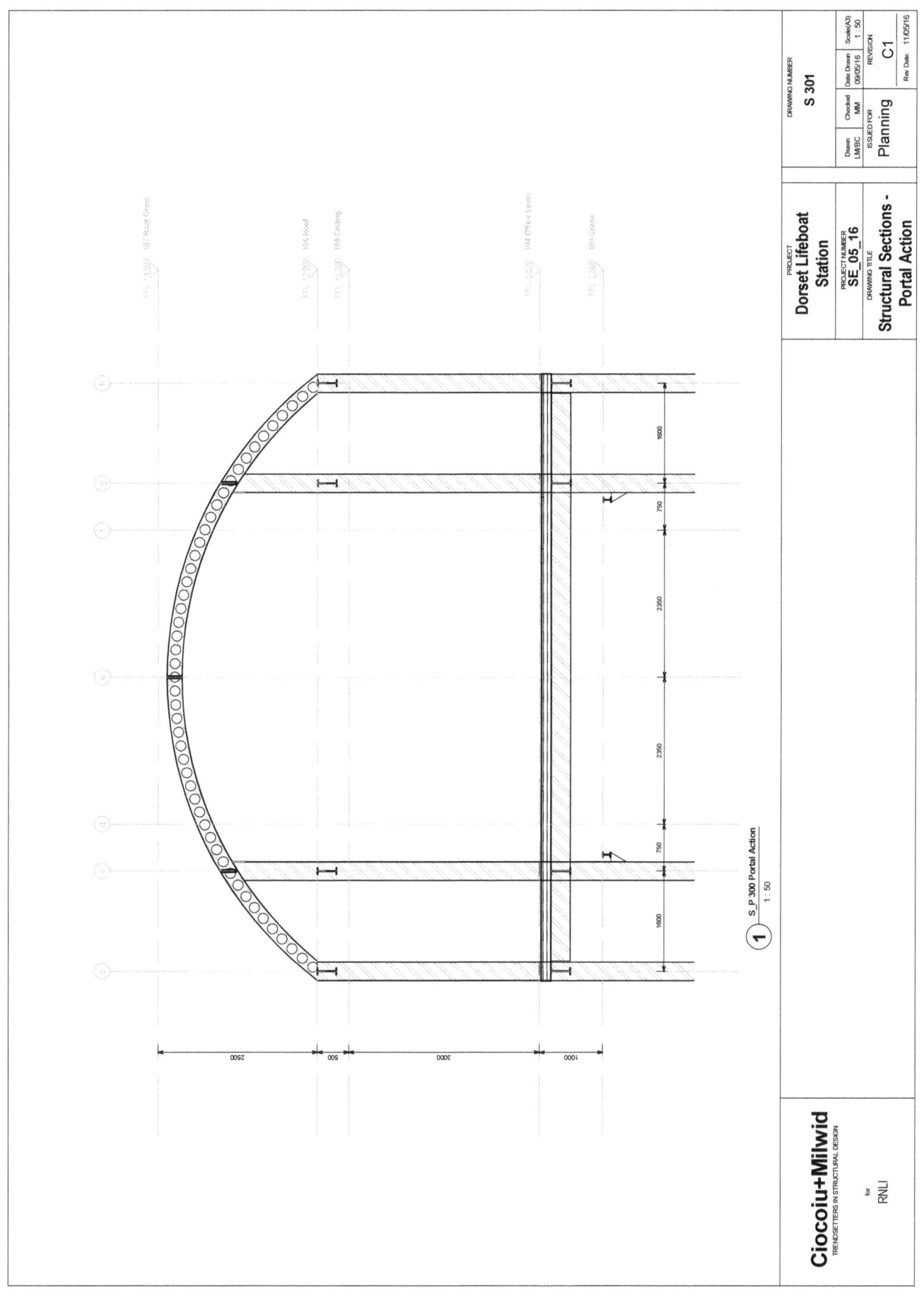

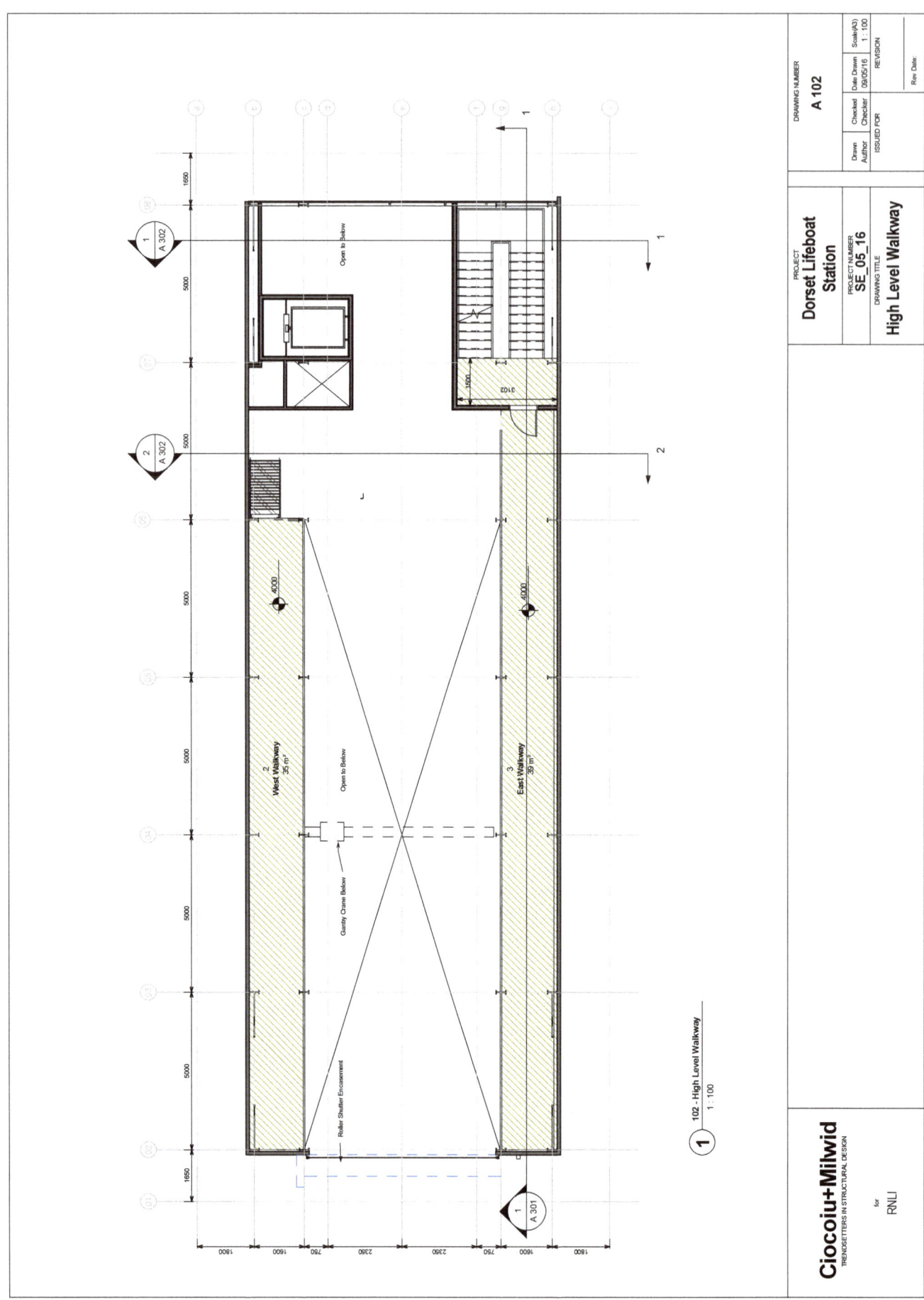

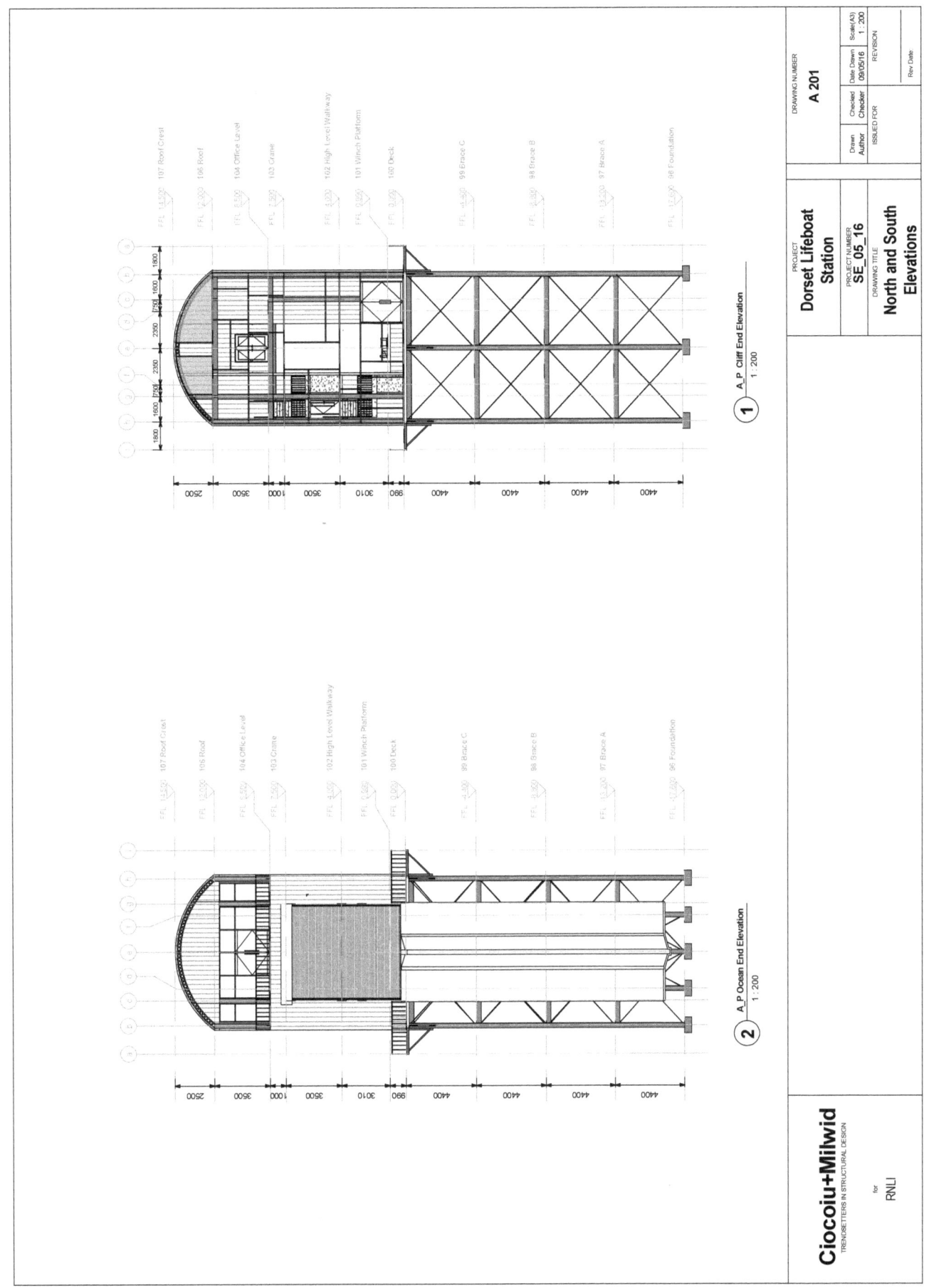

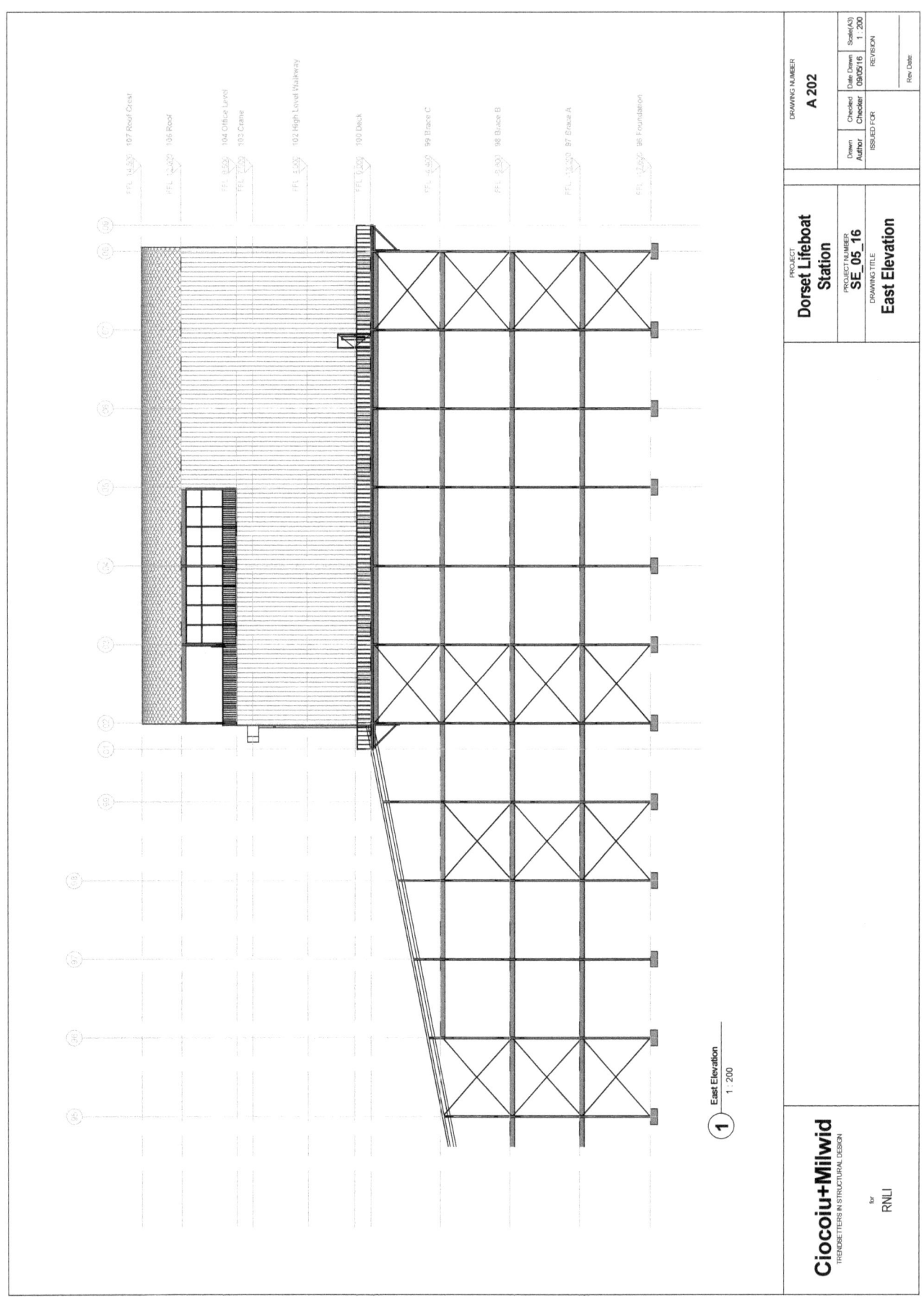

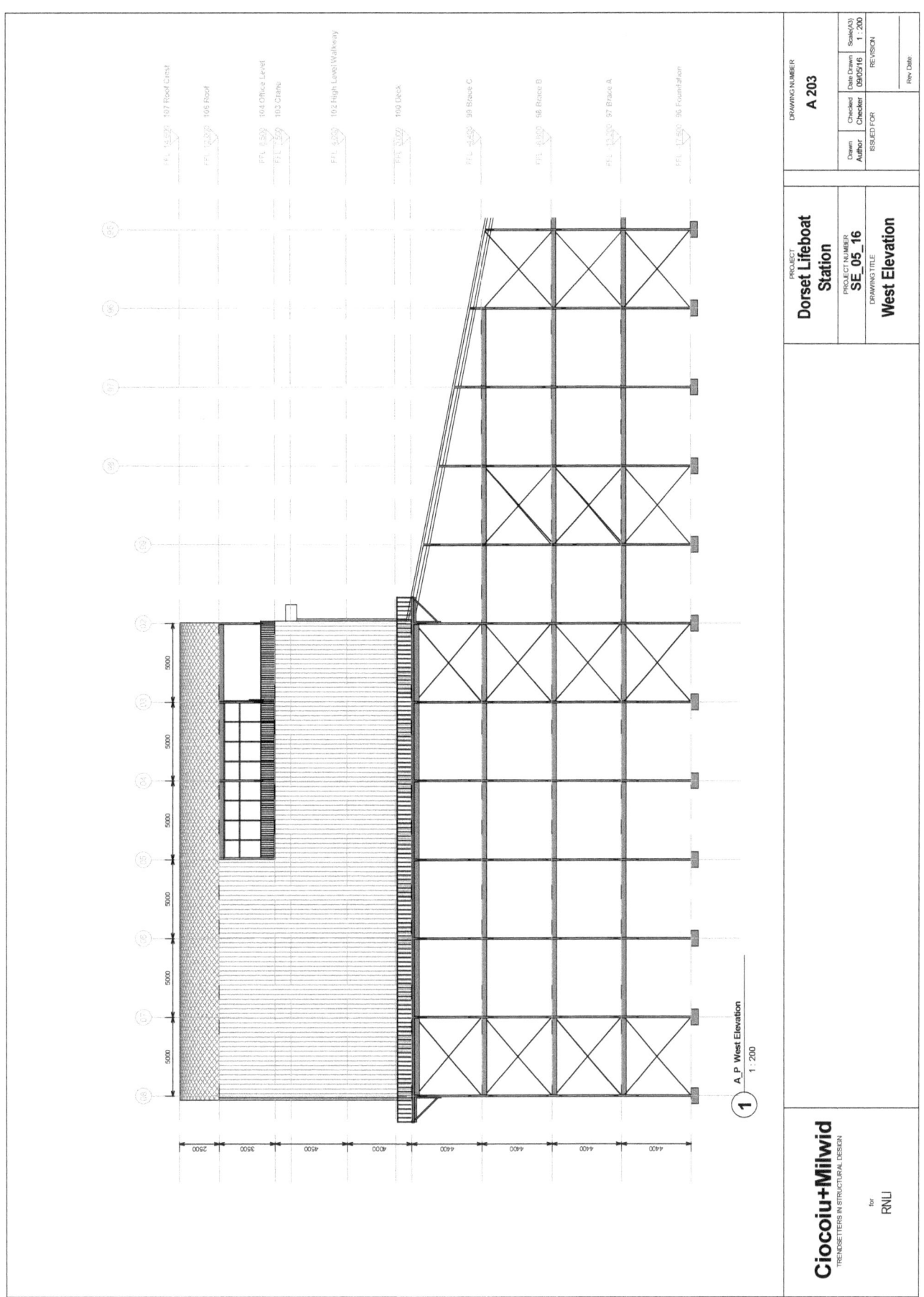

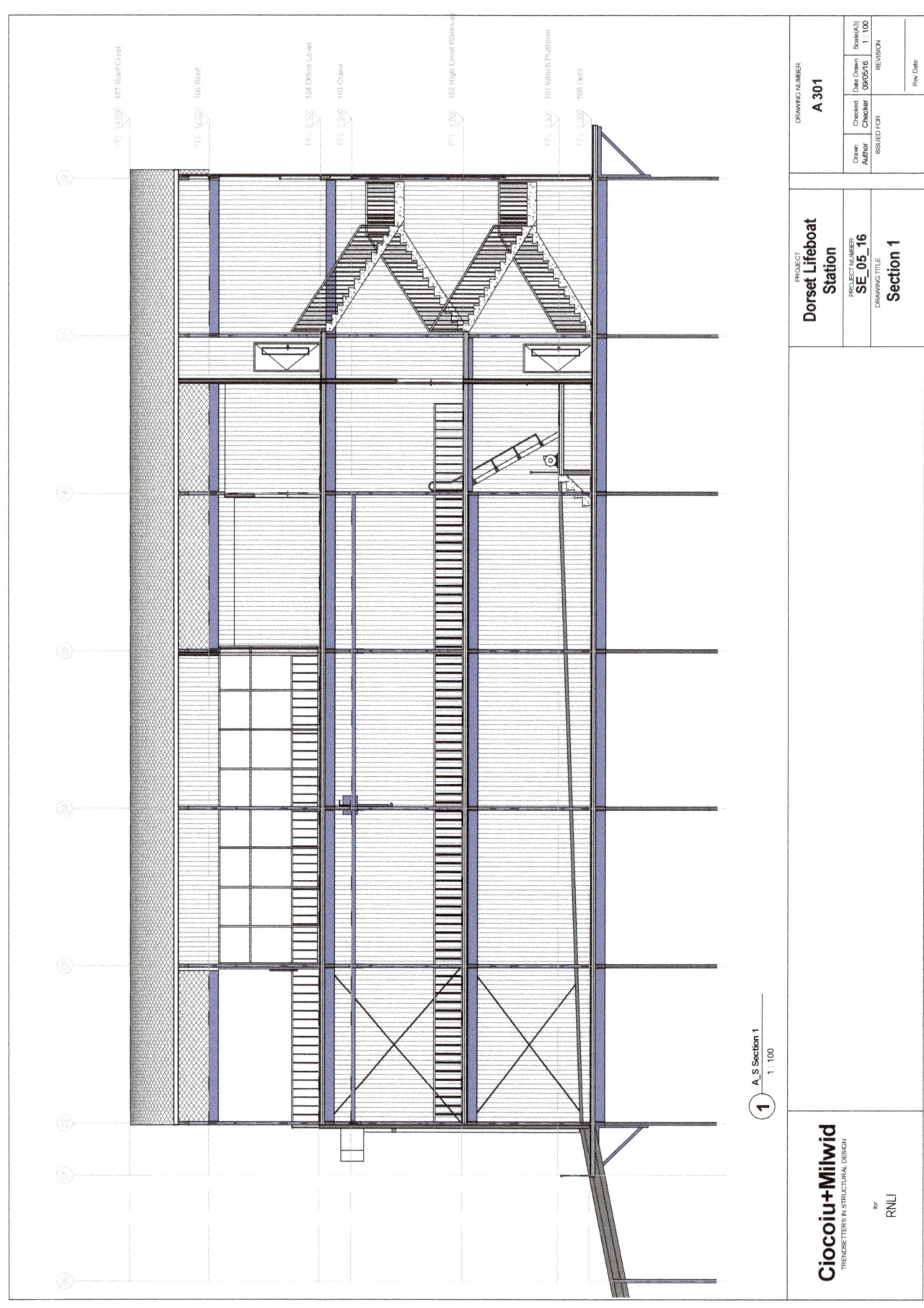

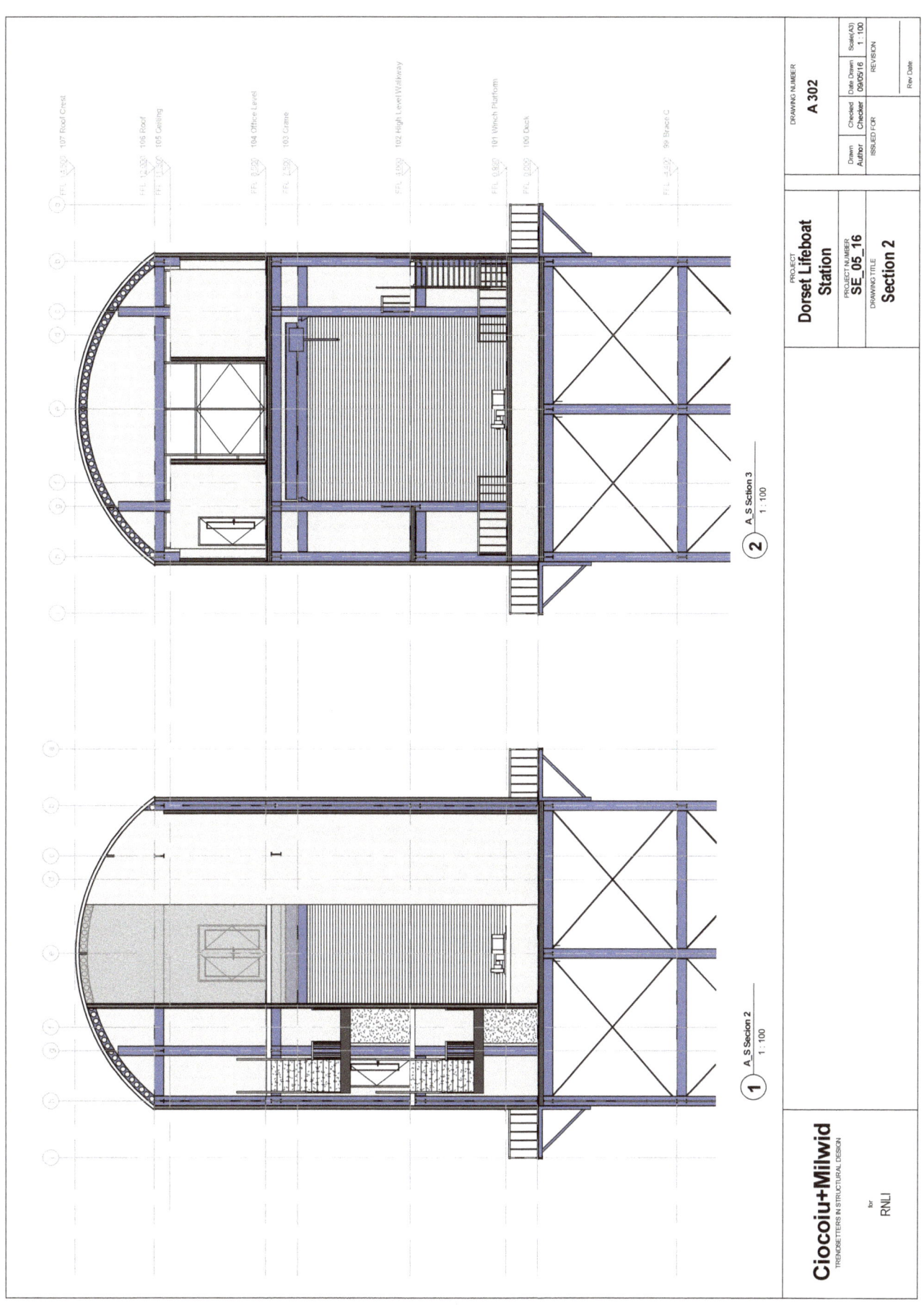

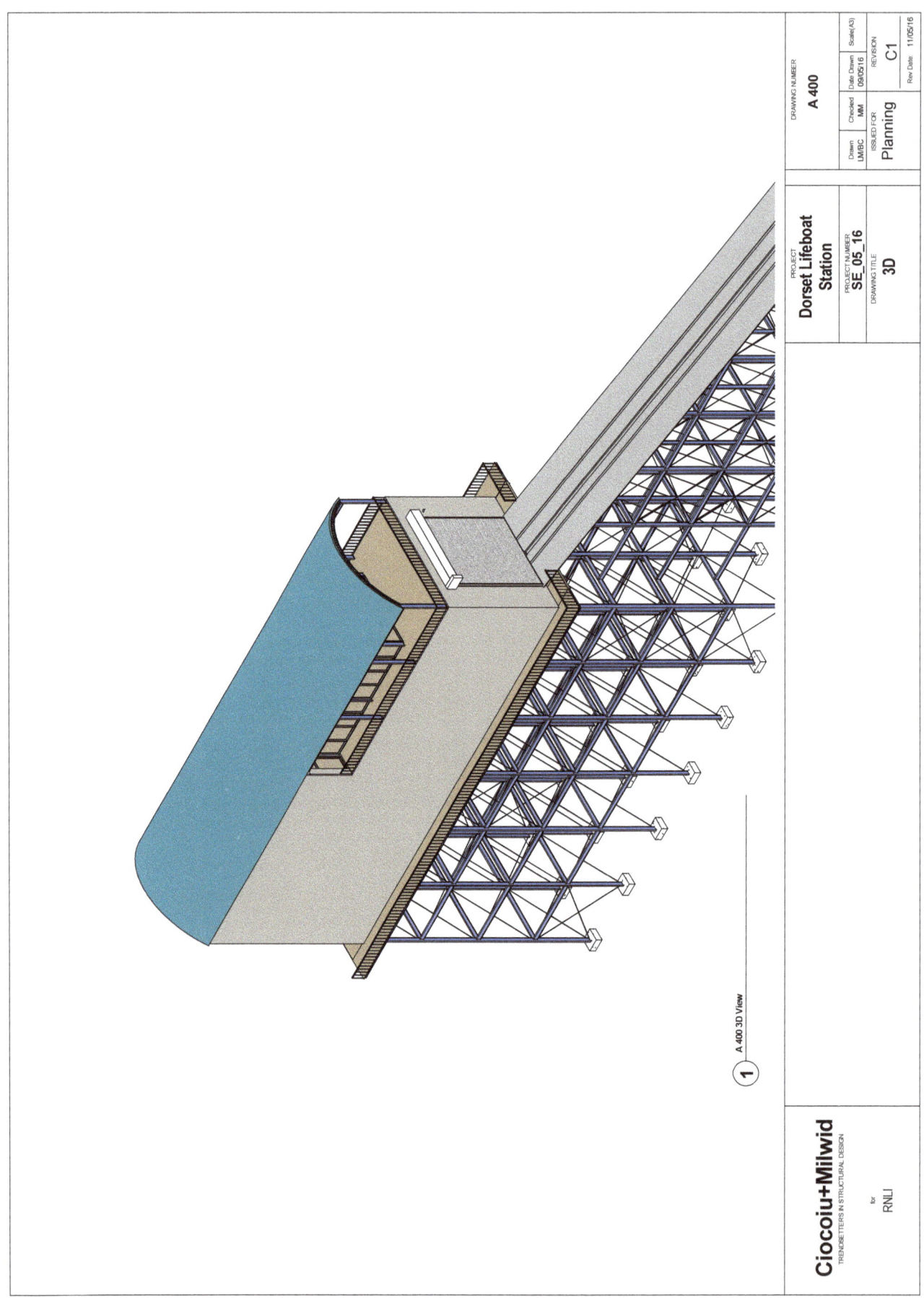

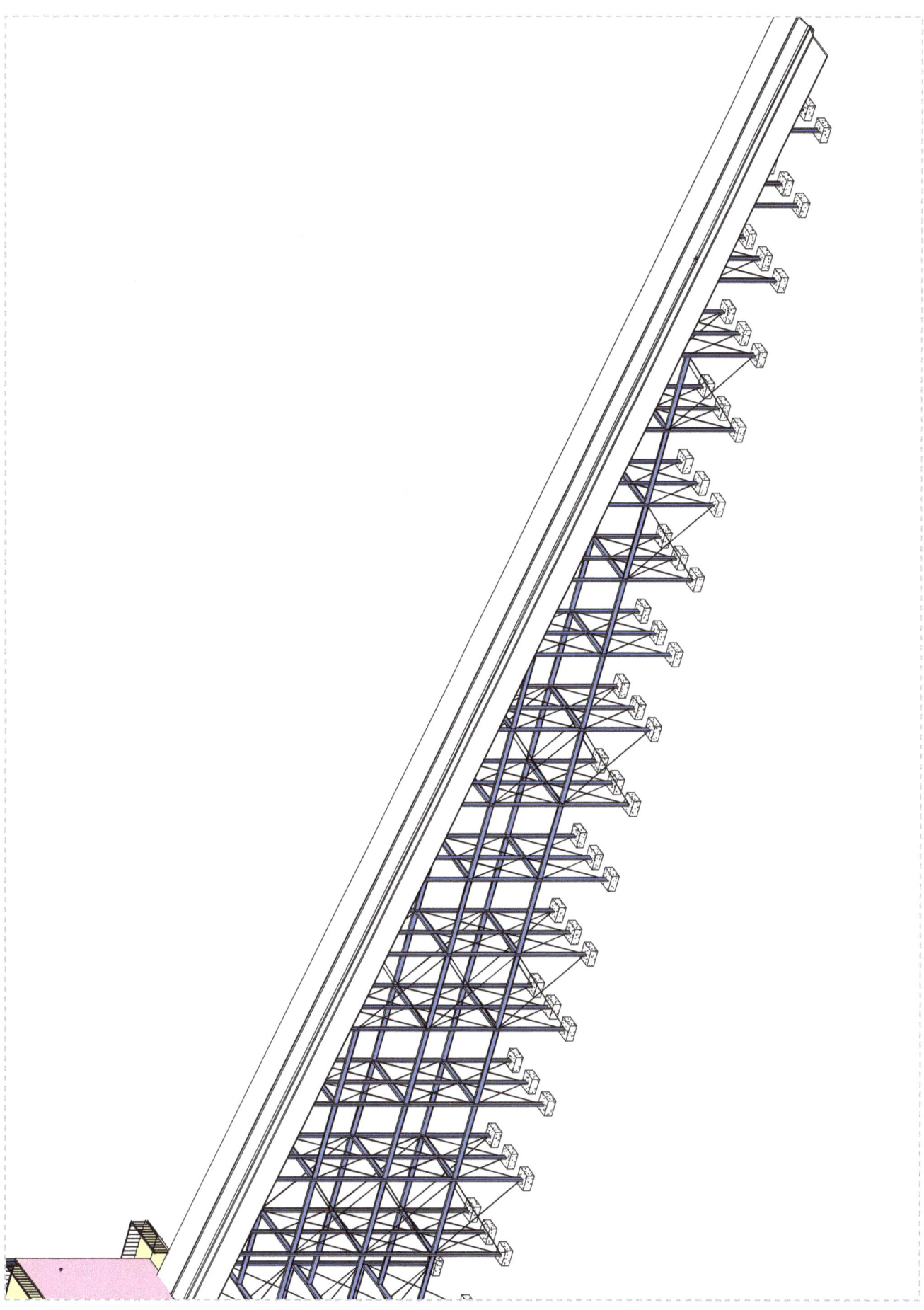

www.ingramcontent.com/pod-product-compliance
Lightning Source LLC
Chambersburg PA
CBHW041545220526
45473CB00014B/2960